Instrument Flying Refresher

Also by Richard L. Collins

Air Crashes
Flying Safely
Flying IFR
Flying the Weather Map
Dick Collins' Tips to Fly By
Thunderstorms and Airplanes
Flight Level Flying
The Perfect Flight
Pilot Upgrade (with Patrick Bradley)
Mastering the Systems

Instrument Flying Refresher

Richard L. Collins
Patrick E. Bradley

An Eleanor Friede Book
THOMASSON-GRANT
Charlottesville, Virginia

To the flight instructor,
often overworked, underpaid,
and en route somewhere else —
but there, ready to fly

Published in 1992 by Thomasson-Grant

Any inquiries should be directed to:
Thomasson-Grant, Inc.
One Morton Drive
Charlottesville, Virginia 22903-6806
(804) 977-1780

Printed in the United States

99 98 97 96 95 94 93 92   5 4 3 2 1

Library of Congress Cataloging-in-Publication Data

Collins, Richard L., 1933-
        Instrument flying refresher / Richard L. Collins, Patrick E.
    Bradley.
            p. cm.
        Originally published: New York : Macmillan ; London :
    Collier Macmillan, ©1984.
        "An Eleanor Friede book."
        Includes index.
        ISBN 1-56566-023-4
        1. Instrument flying. II. Bradley, Patrick E. II. Title.
    TL711.B6C65 1992
    629.132'5214--dc20   92-14751
                                        CIP

Contents

Foreword

No question, there is a lot to instrument flying. From the day we start to study the subject to the day the old hat is hung up (hopefully not on a tree) there is a myriad of important stuff, minutia, and the ever-present requirement to maintain the ability to fly the airplane. All this might seem awesome at first glance, but instrument flying is actually rather easy.

Easy? Instrument flying easy? That might sound preposterous in the beginning. It might not make sense to a beginner, a rusty instrument pilot, or to one who obtained, but never used, a rating. But it is true. Put another way, instrument flying is as hard as you care to make it, but it becomes difficult only when approached as a mechanical exercise. A pilot can make it easy by studying the activity, by practicing to proficiency and staying refreshed, and most of all by enthusiastic participation. Instrument flying is more of a *thinking* exercise than other forms of flying — recognize it as such and things work better.

The purpose of this book is to help with the understanding of instrument flying as a thinking game. It is to refresh by sharing participation. There are a lot of things in instrument flying that we tend to forget — at least we don't think about them often enough. The reminders are here.

Each chapter begins with an overview by a relatively new instrument pilot, Patrick Bradley. His thoughts on the various aspects of instrument flying point out how it looks as a pilot works into instrument flying. I then follow with an expansion on the subject, based as much on flying with and observing new and rusty instrument pilots as on twenty-nine years of IFR operation in light airplanes.

I have flown a lot with Bradley and watched him evolve from a beginner, a novice instrument pilot, to a pilot who understands and knows his way around the system, and who has a measure of weather wisdom.

The day I knew he had the big picture was a stormy one. I was en route to Florida in my Cessna P210; Bradley was supposed to fly a Piper Archer from Teterboro, in New Jersey, to Nashville, Tennessee. My son Richard — another new instrument pilot I watched evolve — was flying my airplane, and we had a wet and bumpy climb from our Trenton, New Jersey, base. Would Patrick go to Nashville? The weather was not severe, but it was of a nature that could and should deter a reluctant instrument pilot. Then I heard him on the frequency. He relates the tale of that flight in this book. I think he is as convinced as I that on that day he discovered his ability to fly IFR and the confidence to use that ability in the interface between weather, the air traffic control system, and a real weather system.

Getting it together as a pilot doesn't mean that weaknesses don't remain. They do. We all have them, and some go away and then come back. The process of learning and refreshing is how we keep these weaknesses from leading to substantial problems.

Accidents are used throughout this book, as in my previous books, as examples. Many of these are airline accidents. They are used because the cockpit voice recorders give a unique insight into the thoughts of a crew as they flew an airplane to the scene of an accident. The intention is to learn from their experience. If we can develop some feeling for what other pilots were thinking before an accident, it better prepares us to handle a similar situation should one arise.

When considering the mistakes of others, we'd all be kidding ourselves if we said that we might not at least start along the same path. It is a lot easier to catch another pilot's mistakes than to catch your own — especially from careful and unhurried study on the ground. It's even easier in the air: There is a big difference between sitting in the left seat and managing the whole thing and sitting in the right seat, charged only with being a critic of another person's technique and decisions. But the beginning mastery of instrument flying comes when we develop the ability to be the equivalent of a right-seat critic of our own left-seat flying. Single-pilot IFR works best when there is a continuous self-evaluation of performance. Adding our experiences to those of others can be a great aid.

Experience is, after all, where real knowledge of flying originates. No one pilot will ever see it all, and one experience might not transfer to another situation. But it is possible to share through study and refresher training.

Richard L. Collins

CHAPTER 1

The Psychology of Instrument Flying

FIRST PART   Patrick E. Bradley

It didn't take long after I got my private pilot's license for me to learn that there were distinct limitations in being banished from the clouds: the vacations that I spent 200 miles from my planned destination because of stalled fronts, the canceled trips and disappointed friends who found it difficult convincing me that they understood how the weather could be fine at our departure point and crummy at our destination. And there were the questions about just how good my private license was when I could never be sure of completing a trip because of weather. Well, before I even had the 200 hours necessary to take the instrument flight test I decided that I would have an instrument rating, and that once I did, I would fly wherever I wanted whenever I wanted. I was ignorant, I know now, but my heart was in the right place, and I have never regretted that uninformed decision.

What I knew about instrument flying would have fit into a fuel strainer, but I did know intuitively that getting an instrument rating wouldn't be a breeze. Money is always a concern, especially considering not just the price of the rating but also the cost of staying current. Time is another problem. How do

1

you find the time to wage a full-scale assault on instrument flying? And then there are the accidents so graphically described in the newspapers — what assurance did I have that I could be not just an instrument pilot but a *capable* one too? I didn't dwell on these questions. It ended up costing me some time and lots of travail. But on the other hand, I went ahead and pursued my goal. I believed that with the proper training and a measure of dedication I could be a capable instrument pilot with the ability to use flying as a tool as well as a pastime.

My biggest problem in beginning my instrument training was that I didn't have enough information to make getting my rating as easy as it should have been. I lacked a game plan, and no one, including my instructors, seemed able to help me formulate one. Although I started from scratch three times with three instructors, not one of them sat me down on the first day and told me what I could expect in terms of work load, setbacks, scheduling, or financing while I worked with them to prepare for my rating. Instructors see their role as instructing, and will seldom confide their own mistakes and obstacles. But knowing where you are going is important; its absence is probably one of the most significant causes of attrition from the ranks of instrument students.

Beyond the logistics of getting the rating, I had to contend with the new skills I was expected to learn. My instructors were generally effective in describing maneuvers, and I was generally adept at picking them up. I learned that instrument flying consisted simply of being able to perform particular maneuvers or procedures without looking outside the airplane. The first time I flew on an instrument flight plan by myself I learned there was much more.

Wisely, I ventured up on a beautiful day, not a cloud in sight, and no hood screwed onto my head. I just wanted to get accustomed to the instrument system and to talk to the controllers as the "big boys" did. I got the clearance straight, and read it back like a pro. I got the airplane off the ground, climbed to my altitude, and got nicely squared away on my airway. But then much to my surprise and chagrin, I flew out of radar coverage. Things started going down the tube in a hurry.

CONTROLLER: Ah, 6049J, what's your present position?
ME: Um, stand by a second.
CONTROLLER: Roger.
ME: Approach? This is 49J and uh, I'm ah, just south of SILKY intersection.
CONTROLLER: Roger 49J. Ah, can you do any better than that?

At this point I realized that I really didn't know where I was. I was reasonably sure that I was on my airway south of SILKY intersection, but how far south or when I could expect to reach the point at which I would have to begin my transition to the approach was a mystery. After a few minutes I calculated that I was 5 miles from SILKY on V-293 at 3,000 feet, but these are things I ought to have known all along. Even after a lot of IFR flying, I'm still caught unawares at times; nonetheless, the realization that I have to know my position is reinforced on every IFR flight.

Knowing one's position is essential through every phase of an IFR flight, and it is one of the things that most often gets me into trouble — especially on the approach. One of the first things I learned when I began to fly in the clouds was that losing my position was a greater problem than losing my balance. Of course, the penalty for spatial disorientation is usually more severe than that for positional disorientation. But spatial disorientation is a matter of technical skills whereas positional orientation requires constant mental activity and a clear knowledge of how your navigation instruments can work to tell you where you are and where you're going.

A favorite exercise of one of my instructors was to shoot a low approach and then have the controller vector me for another approach — I would never know which approach or for which runway. After about five minutes of vectoring and climbing and cleaning up the airplane, my instructor would ask, "Okay, now show me where we are on the approach plate." At that point I would generally go through the manual procedure of locating my position on the approach plate, and would end up getting behind on some other aspect of the upcoming approach. I suppose that what he was trying to teach me was to incorporate positional awareness into my thought processes. It not only included being able to locate my position mechanically, but also constantly being aware of where I was going and verifying my path with instrument indications. Instrument trends became important in the approach phase, an ADF became my best friend, and a VOR, if on the field, became pretty much useless.

Familiarity with instrument procedures, or rather the lack thereof, was a prime factor in my inability to, say, stay abreast of my position, or to stay ahead of the airplane during an approach, or it would cause me to just plain forget some important element. I didn't have a real grasp of what to think of and when, and I'm still struggling with that today. But I know how important it is to learn the basis for what things to do and when to do them. These are the skills that make an instrument pilot an expert. To a large extent, learning to think about the right things at the right time comes with experience, but it is also an important element of training. I often found that too much of my training concentrated on how to execute the mechanics of flying on instruments instead of on the much more subtle skills of how to fly intelligently.

But how can you prepare someone for every eventuality that they will face while flying? I guess you can't, really. That's why instruction I valued the most was that which taught me to think, and to think under pressure. With it, I was able to get a feel for the telltale signs of a difficult approach. I learned the questions to ask controllers if I began to think my margin for error was getting too tight. I also learned what my priorities ought to be when I just couldn't do everything at once. The best pilots, I think, aren't the ones with three hands; they are the ones who know what to do with the two that they've got. These are the qualities that make up a good pilot, along with perhaps one more: a healthy respect for the weather and the airplane, and a knowledge that left uncontrolled or unheeded, warning signals can lead to the untimely downfall even of very skilled but perhaps unthinking pilots.

But instrument flying doesn't take exceptional intelligence, nor is it an uncomfortable strain. IFR does not imply constant peril, nor is one even uncomfortable after becoming accustomed to the sensation of not looking outside of the airplane. I was amazed, the first time I flew "hard IFR," how simple and even relaxing it was. I found that I settled into a groove, taking in the information and keeping up with the chores that applied to the phase of the flight. I approached the task with the comforting sense that I was putting my skills to work while improving them more effectively than I could under a hood. Hard IFR is at once relaxing and challenging. Sometimes, on long legs, it can be boring, and on some approaches it can be hair-raising. But for the prepared IFR pilot, hard IFR in actual conditions is the natural outgrowth of the skills that he has gone to some difficulty to acquire and to develop. It is the thing that makes the practice and the studying worthwhile. For some of us it is the reward in itself. Getting where you are going with a modicum of assuredness is the added bonus.

SECOND PART   Richard L. Collins

Like Bradley, most pilots learn visual flying first and add instrument flying capability later on. This in itself is a big factor when a pilot has difficulty grasping the fine points of instrument flying. The VFR and IFR flying systems are very different, and the fact that VFR in most cases is first learned in isolation, with the total exclusion of IFR, adds an air of mystery to the excluded activity. The longer a pilot waits to tackle instrument flying, the more difficult it will appear. Old-timer VFR pilots have an especially tough time when and if they start on IFR — not because they aren't good pilots but because it's difficult to open the mind to something that has been excluded for an extended period of time.

The same thing is true of an instrument-rated pilot who is just starting to use the rating. It's possible to earn the ticket in a sterile atmosphere. Many do it with no actual cloud flying, a minimum of exposure to the air traffic control system, and nothing more than airwork, flying holding patterns, and shooting approaches. And shooting approaches. And shooting approaches. And when you watch the product of "shooting approaches" operate in the system, and discover things like real honest-to-goodness missed approaches, in clouds yet, you quickly learn that it's possible to get an instrument rating without getting the big picture.

There was an accident a while back in which two very rusty instrument-rated pilots (the one flying didn't come close to the

FAA's skimpy recent experience requirements) came a cropper after missing a few approaches to an airport that was right at minimums. It was nighttime, the airplane didn't have a glide-slope, and the pilot failed to activate the pilot-controlled approach lights at the airport. In reading between the lines, you couldn't help but have the strong feeling that there wasn't a lot of knowledge on board about what works, what doesn't work, and what really counts in instrument flying. These pilots were getting a "refresher" at the toughest possible time.

WHAT DOES COUNT?

A little guessing game: What pilot attribute is the most critical in instrument flying? Is it hand-eye coordination? The ability to react rapidly? Excellence at shooting approaches? Flawless holding pattern entries? None of the above? The right answer is the last one. The most desirable virtue for the instrument pilot is the ability to think and reason.

Instrument flying is a thinking game; having the correct information stored and keeping the mind very active makes it work. Having the information is primary because it is difficult to do a proper job of thinking when there is a question or problem in mind. Flying and navigating the airplane on instruments requires a continuous process of interpretation and decision making. If there is fixation on any one thing — and we do fixate on questions or problems — everything else suffers. I've often thought that a metronome would help some pilots do a better job of instrument flying. Each click would be the key to move the eyes and the mind on to another of the items begging for attention. I've seen excellent results time and again after getting a pilot to talk aloud to himself about the task at hand during a problem portion of instrument flying. Talking stimulates the mind. To have something to say, you have to keep looking at different things to get new information.

It's possible, too, for a pilot to grade his store of information as well as his ability to think. You have only to review flights carefully — make notes in flight if you will — and record the number of times that questions interfered with the thought process. Then you know which skills need refreshing.

QUIZZICAL EXAMPLE

I was flying to Kerrville, Texas, one grungy day and was cleared direct to the outer compass locator from a good distance north of Kerrville. The vortac is west of the locator, and the distance from the vortac to the locator is 10.8 miles. The only published route on the approach plate is from the northeast.

Even though I wasn't flying it, I fixated on the published route. I must have done this simply because it was on the chart. As the DME counted down to 10.8 I thought that I'd pass the locator. But the ADF needle remained rock steady, straight ahead, and there was nary a peep from the marker receiver.

Hmmmmm. The old brain started grinding to a halt, stuck on one subject. Why hadn't I passed the marker if the DME had counted down below 10.8? The answer wasn't long coming. The DME kept counting down, to about nine, and then started counting back up. When it got back to 10.8 I crossed the locator. If I had been on the published route, it would have been a countdown all the way, but coming in from the north I flew closer than 10.8 to the vortac en route to the locator. I gave myself a poor grade for that part of the flight, and the moment of confusion probably caused the approach to work less well than I would have liked. I made a mental note to do a better job of orientation.

The thinking process can suffer when there is confusion about word from the ground. An aspiring instrument pilot was flying my airplane in actual conditions. We were at Flight Level 190, and as we neared the destination, the controller called with a clearance to "cross 10 south of Salisbury at 17,000."

The pilot turned to me with a totally blank look on his face. His training had all been at low altitudes; he had never heard a clearance to cross a point at an altitude. The microphone might as well have been red hot; he wanted mightily to get rid of it. There was no way he could read back a clearance that he didn't understand. A few minutes later we got another "cleared to cross" message, and he could handle it that time only after being told exactly what to say. The introduction of mystery — at least it was a mystery to him — had partially paralyzed his brain.

THE TEETH

The thought processes one has when flying instruments in rainy turbulence have always been interesting to me. At such a time I feel alone, and in truth, there's probably no situation in which you depend as much on yourself. If you mishandle your sailboat and break a mast, there's always the life preserver. If you mishandle your car and spin out, there's always the belt. But when the big sky in which we fly our little ships turns surly and starts churning, it's one brain against the elements.

What do I think about? I have a plan. First, I remind myself that I studied the weather situation carefully enough and used the thunderstorm avoidance tools at my disposal carefully enough, and that any situation I get into will be manageable if not comfortable. Once you are in turbulence, you have to have confidence. I think a lot of airplanes are lost simply because a terrified pilot's brain ceases functioning and he gives up. On the other hand, if the old brain offers moral support and keeps grinding out the interpretations of what's needed, things should work okay.

The next thing I usually think about is the structure of the airplane. I have a lot of friends who are aeronautical engineers and design airplanes; while they often come up a little short on systems, they do a good job on the structure of the airplane. A plane seldom breaks unless the pilot flies it into a *severe* storm, or unless the pilot overstresses it. That's my job. I worked at staying out of the worst part of the weather, now I'm working at minimizing the pounding that the airframe is taking. With all the assets counted and the challenge defined, the concentration is on flying.

WHERE AM I?

The term *positional awareness* is useful in describing the requirement to always know where you are in an airplane. If the question Where am I? is asked and answered continuously and correctly during a flight, and if the location you describe to yourself is a correct one for that time in the flight, that's good. Bad things will not happen. And the flying will be easy, or at least easier.

Continuous questioning and awareness of position can even

help make a tolerable situation out of one that might otherwise deteriorate into a difficult time.

This was brought home to me on an approach to the Vero Beach, Florida, airport one very stormy day. There was a lot of mean weather about. Florida had twenty-six tornadoes that day. My departure from Melbourne, a short distance north of Vero Beach, was delayed for a few minutes to allow a vigorous storm to pass the field. When I took off, the picture on the aircraft radar as well as on the Stormscope suggested that I'd have to stay a little west of a direct line to Vero. This was fine with air traffic control.

When I was about 10 miles north of Vero I talked with the tower there, and the controller confirmed what I thought I saw on my radar. He said it was raining so hard just west of the field that the sky was green. That prompted me to switch back to the air route traffic control center and ask him if I could orbit awhile and wait for the williwaw to clear the airport. That was fine, but here I collected a few demerits in the thinking department. I was so thoroughly engrossed in the process of evading the weather that I wasn't doing a very precise job of flying the airplane. I was 100 feet low, and the airspeed was 15 knots under where I wanted it because I hadn't pushed the power back up enough after a descent. A little mental kick in the pants got things back in shape.

I did get a reasonable grade on awareness of position when shooting the approach a few minutes later. The controller was vectoring me to the final approach course, and he kept adding degrees. The inflow into the storm was very strong, and I later calculated that there had been 40 degrees of drift to the left as I flew the ever-increasing vector headings to the final approach course. And although the controller had been aiming for interception 7 miles from the vortac, the actual point of interception was very close to the station. I was able to psyche myself up for this, though, and to get ready for the 100-degree turn to the left that would put me on final only a mile from the vortac.

REASONING

There was a lot more than awareness of position involved in the thinking process that morning. A plan had to be formulated and stored for recall in case of a missed approach. As the aircraft's altitude (vertical position) changed in the approach, wind shear that would require pilot action was highly probable. I had to try and visualize a cross section of the air, and the airplane's relationship to the changing direction and velocity. It would be a circling approach. Originally I had planned to break to the left for a right circle, but I descended out of the clouds too late for that and had to circle left, which meant a new plan. I did anticipate the tremendous sinking spell as I turned and descended into a stronger tailwind. When the airplane developed that old falling feeling, a good dose of maximum continuous power kept it going and at the correct height.

The last thinking demerit was collected on final approach. I elected to land with approach flaps because of the strong surface wind, but I never got around to calculating a proper approach speed. It would have been about 90 knots, allowing an extra 10 for gusts. I burned in there at 105, floated, landed long, and taxied in with smoking brakes. I had a slight headache after parking the airplane — a product of overheating the bearings in my noggin. The thought processes on that short 25-minute flight were as involved as those on an average flight covering half the country. And, wouldn't you know it, even though I was on time, the people I was meeting for lunch had already gone because they didn't think I'd show up!

THERE'LL BE A TEST

It is important, educational, and even enjoyable to examine thought processes after a flight. It's part of a continuous refresher on instrument flying. On the flight described, I could come up with the reason for a little sloppy flying while I was holding, but I honestly don't know why I flew the final approach 15 knots too fast. Maybe it just felt better in the turbulence. The refresher key is in filing that bit of information. I promise to remember next time and fly the airplane at the correct speed.

DIVERSIONS

Fighting diversions is an important part of the thinking business. When I first started flying instruments I had a terrible time keeping my eyes on the panel and off the windshield in heavy rain. The sound of the rain would draw my attention; I'd catch myself staring intently at the water streaming back. The clue to the result of my fixation was often an increase in sound level as the airplane started going its own way because I wasn't minding the store. The same thing was true when flying in clouds of varying texture. When it was opaque white out there, my vision was never drawn to the outside. But with

changing cloud density, and thus changing light, I had to mentally horsewhip myself to do a proper job of staying on the
gauges.

WHERE TO LOOK

Whenever flying isn't going well, it's time to analyze everything being done. Think it all the way through. A person
might be the world's best natural pilot, but if he doesn't look
at the right thing at the right time, everything will go astray.
And just as we have to keep up with the location of the airplane,
we have to keep up with the location of our thoughts.

One day I was trying to analyze why another pilot was
having a little wrestling match on an ILS approach. In watching the relationship between eye movements and control movements, I finally figured out that he was moving the controls
when he was looking at the navigation needles. His thinking
process was out of phase. It had him doing something that
doesn't work. It was a simple matter of his looking at the
wrong thing when he was changing the attitude of the airplane,
resulting in overcontrolling. He knew better; he knew he should
look at the attitude indicator, but he had a lapse — like the
one I had when flying the 105-knot approach.

THE BIG PICTURE

Much is made of having "the big picture," or, as they say
in governmentspeak, an "overview." And it is *very* important

in instrument flying. Without the big picture a pilot will often find himself doing something illogical.

For example, we all know that the wind-aloft forecasts are, at best, computerized guesses. But they are part of the big picture and shouldn't be totally ignored.

One day, fleeing to Florida from New Jersey in the wintertime, I watched another pilot fly a heading of 265 degrees for a while, after being cleared direct to Waterloo, a vortac quite a distance ahead. The bearing to Waterloo was 200 degrees. I finally asked the pilot why he was flying 265 when Florida was down thataway, to the left. He replied that the navigation needle was off to the right a little and was slow coming in. I agreed with that, but disagreed that with a forecast 30-knot wind we needed 55 degrees of correction to intercept the radial. Big picture: You get the airplane to where it is going by pointing it in the correct direction: Calculate the amount of expected wind drift expected and be highly suspicious of any heading very far from that calculated. If it's off much, it means either the airplane isn't tracking toward the destination, or, the wind-aloft forecast is in error and a different heading will be required to make the airplane go where you want to go.

LIABILITIES

If we study instrument pilots and decide that an ability to think about the right things at the right time is a big asset, what might be the biggest liability? What makes a dangerous instrument pilot?

The accident records tell a very clear story, and illustrate that the things that bug us when flying — confusing clearances and the like — are not the things that lead to real problems in instrument flying. We also see that experience is not a strong factor in preventing accidents. There are interesting differences in the accidents that occur in singles and twins, and there is clear evidence that weather-avoidance equipment helps with risk management.

Of ninety total-IFR accidents examined (about one year's experience), only seven involved pilots with less than 500 hours total flying time. All these relatively low-time pilots were flying single-engine airplanes. Of these, two were pilots without instrument ratings operating on IFR flight plans (one had only about 50 hours total time) so two don't really count as legitimate IFR accidents. Of the five that do count, one pilot lost control of the aircraft on takeoff and nobody got hurt. On another, the pilot mismanaged the fuel, resulting in a forced landing. One involved personal injury — the airplane hit trees while flying at an altitude below minimums and was then successfully landed. Two of the seven were fatal accidents. Both involved high-performance singles being operated at night in an area of thunderstorms. In both cases, the airframe (one a Cessna 210, one a Piper PA-32) failed in flight. You could say that flying experience, or any lack of it, would change the thunderstorm exposure, but I don't think it would be a great factor.

Past those two serious accidents, low-time instrument pilots had few problems in this time period. So, you can't say that the beginning stands out as a dangerous time in a pilot's IFR career. Perhaps new instrument pilots are more cautious, plan more carefully, avoid the temptations to carry on regardless of weather, and are simply more current because of recent training.

Still, looking at single-engine airplanes, we see the greatest accident involvement by pilots with from 500 to 1,500 hours total time. The greatest number of accidents are related to descending below minimums, sniffing for asphalt you might say. Pilots of the 500- to 1,500-hour experience level are great participants in this type of accident. They also tangle with ice and thunderstorms, a substantial part of the single-engine fatal IFR accident picture in this time period. Sixteen percent of the fatals were ice related and 21 percent involved thunderstorms. For pilots flying singles, management of these two items is a big part of avoiding the "dangerous" pilot label.

(There's an interesting twist to the statistic on descending below minimums and flying into the ground. Fatalities are involved in only half the single-engine accidents where the pilot descended below minimums and hit something other than the runway. Almost 90 percent of the same-type accidents were fatal in multiengine airplanes, and the ones that were not fatal were in very light twins, so it's clear that the lighter and slower the airplane the more survivable the accident.)

System malfunctions claim as many single-engine pilots as ice. This is a clear call for a pilot to minimize risks through continuous refresher training in the procedures to use following systems failures (mainly partial-panel flying), and in equipping the airplane so the failure of any one system won't leave the pilot in a situation that he's not equipped to handle.

TWINS

Looking at general aviation multiengine airplanes that are involved in IFR accidents, we see a somewhat different picture. The pilot is much more experienced — the greatest flying time concentration is between 3,000 and 8,000 hours, showing that a fat logbook certainly isn't a barrier against problems. Going below minimums is the primary problem in twins. Looking down the list you see some fancy airplanes — including a lot of pressurized piston and turboprop airplanes. As noted earlier, this type of accident is much more likely to be fatal in a larger, heavier airplane than in a lighter one. In either case, single or twin, there is no question that a pilot who doesn't slavishly adhere to minimum altitudes and proper approach procedures is dangerous regardless of what airplane he flies and how much experience he has. Night is a big factor in approach accidents. Over half occur in the dark; probably only about 10 percent of the approaches are shot at night.

Ice is another twin pilot's problem. Ice was a factor in 16 percent of the total twin IFR accidents, and all the airplanes involved were types that usually are equipped for flight in known icing conditions. I think the strong possibility exists that pilots rely too much on equipment and not enough on avoidance of ice even though allegedly equipped to fly in it. Ice can accumulate on unprotected surfaces and result in substantial performance deterioration even on fully deiced machines. It is significant that even though ice is a problem in twins, there were half as many fatal ice accidents as in singles and there were no thunderstorm accidents in twins.

In studying the accident records one can add other information and develop a picture on experience and equipment versus accidents. In some past research done for *Flying* magazine, it was determined that piston twins fly twice as many total IFR hours as singles, and that turboprops fly almost twice as many as singles. Also determined was the fact that the mechanical failure of the engine on a single is seldom an identifiable fatal IFR accident cause. Honest engine failure is extremely rare, occurring only on a random basis. There wasn't one in the year used as an example for this discussion. (There were two fatal accidents involving engine failure in twins in the period.) So any difference in singles and twins is more related to pilots, equipment, and airplane systems.

Using the *Flying* research as a basis for exposure, we find that the safety record in piston twin IFR flying is more than twice as good as in singles and that the record in turboprops is four times better. (IFR accidents in pure jet aircraft occur seldom enough that a one-year sample can't be used for any meaningful comparison.) When compared with piston and turboprop twins, the singles have a near monopoly on accidents related to thunderstorms and system failure. Both types of accidents are equipment related and both can be solved by the single-engine pilot through a combination of proficiency maintenance, the purchase of weather avoidance gear (which is much more common on twins but available for singles), and the purchase of redundant or standby systems for the airplane. Ice protection is also available on some singles. The record makes it clear that the pilot who doesn't take steps to protect himself in these areas is in more peril than the pilot who does. When you take care of those things, the single, with its less experienced pilots, has a record almost as good as the piston twin flown by more experienced pilots.

Reading deeper into the accident reports one discovers that there are some clearly defined problems with pilots of both singles and twins. I wouldn't go so far as to call them psychological problems, but some might attach that label.

BELOW MINIMUMS

There is a high incidence of pilots attempting instrument approaches in below-minimum weather. This in itself isn't bad. There are times when the reported weather at the airport is worse than the weather in the approach area, and we've worked for years to preserve the pilot's right to have a look.

The problem comes when the pilot doesn't believe what he sees and presses on into an apparently impossible situation, descending below a safe altitude without visual contact with the runway. There are occasional accidents where the pilot is caught using a homemade instrument approach, which is most certainly looking for trouble. Trying too hard to complete an impossible approach is lethal indeed.

In thunderstorm and ice accidents, the pilot usually presses on past clear warning signals. This is an area where we have to carefully review and refresh our knowledge of the phenomena and question our actions as we fly along.

In thunderstorm areas, you are either in trouble or out of trouble. When you fly into one, you are totally in it. Either it will cause you to lose control, it will break your airplane, or, if you provide the airmanship, it might allow you to trespass and emerge battered but unbroken. The challenge is to stay out. This is difficult to do if you fly IFR in areas where there are a lot of thunderstorms and you don't have weather avoidance gear. To avoid problems, a pilot has to be willing to give a wide berth to all areas of thunderstorms.

Ice is different. It usually builds slowly, and where a pilot in a thunderstorm is in a sudden situation, ice is more like Chinese water torture; it is usually slow but relentless in it's disfiguring decoration of the airplane.

We'll examine accident causes in detail in chapters on the specific phases of flight.

JUDGMENT

Some attention has been given to the teaching of judgment to pilots, and this relates very directly to the question of what makes a "dangerous" pilot. Certainly if a pilot is deemed to have bad judgment, you wouldn't call him a "safe" pilot.

There are a lot more questions than answers on the subject of pilot judgment. To begin, I think that judgment is somewhat like the "right stuff." Either you develop the ability to direct yourself in the more risk-free directions by the time you reach puberty, or you don't develop that ability. You can't teach this to a person later in life, nor do I think that you can assess a person's potential for good or evil in an airplane based solely on observations taken in other areas.

I've known people who live with reckless abandon except when they are flying. There they become paragons of conservatism. They take foolish chances in their business and personal lives. They very likely tempt the fates by driving after they have had more to drink than is proper. Yet in the airplane it's more than eight hours from bottle to throttle, checklists and published procedures all the way.

On the other hand, I know people who live the meticulously conservative life except when flying, where they become nothing short of a wildebeest of the airways. No proficiency flying, lousy maintenance on the airplane, no checklists, no procedures, scud-running — the lot. It has always been interesting to me that the latter are usually the best stick and rudder pilots, the ones with a natural feel of the airplane.

MACHO

There's some thought that the macho image isn't good in an airplane, but I doubt that there's any accident history to back this up. The airplanes that would attract the macho types — the Beech Duke is one — just don't have a bad accident record. And even the enormously ego-satisfying Gates Learjet, which has an overall unfortunate accident history, does very well in the hands of pilots who have been through the factory training program. I think this only suggests that any pilot, macho or otherwise, who realizes that good training and proficiency maintenance is part of the act, is far less likely to fall in the "dangerous" pilot category.

Instead of judgment, it boils down more to a pilot understanding the risks and then managing the risks. Call that good judgment if you wish; it seems more to me like common sense. It means going beyond what the FAA and some instructors teach — simply because they don't understand the real risks.

One other subject I'd like to touch on here is "hard" IFR.

You often hear pilots say that they have an instrument rating but don't fly hard IFR. "Why, I just use it to climb on top of overcasts, or drone along in smooth stratus, and to shoot approaches when the weather is relatively good." Some might say that such a pilot is exhibiting good judgment; actually it is one of the highest forms of self-deception. If that pilot uses the rating at all, he's going to be cruelly surprised and will be, in fact, a very dangerous pilot.

The weather information and forecasting system just isn't up to the task of defining weather precisely enough for a pilot to decide that he's going to take only small nibbles of instrument flying. Anyone who has flown IFR much knows that tops are often much higher than forecast; they build during the day, and the visibility on top can be obscured with higher clouds around. What looks like smooth stratus can often be turbulent because of wind shear, and the good forecast for arrival often folds. A pilot who would attempt to avoid hard IFR based on forecasts can find himself with nothing left but an alternate at minimums and an area of thunderstorms between here and there. Unless the flying is limited to a relatively small area you can get some big surprises. Even in a small area, you can find surprises. Staying sharp is the key.

The other great hard IFR deception comes when a pilot says that he'll fly it in a twin but not in a single. The fact is, virtually all IFR accidents relate only to the relationship between the weather, the airplane and its equipment, and the pilot. The number of engines is not a factor and, in fact, if engines are factored into the equation, it is the twin that comes out with the worst engine-related safety record. To be sure, there are equipment-related items, but these can be easily managed in a single. For example, I won't fly IFR in a single without my dry-battery-powered transceiver to use in case all else fails. And I'm a lot more conservative about thunderstorms when an airplane doesn't have radar or a Stormscope. But the number of engines bolted to the airframe matters not.

In some cases, it's likely that a pilot's outlook on singles and twins contributes to an accident. If we build in our mind this picture of one airplane being capable and another incapable, then there's strong likelihood of our doing things in the "capable" one that we could cull in the "incapable." And, it has been proven time and time again, when you start doing things in a twin that you wouldn't do in a single, you are scouting for trouble.

The redundant systems of the twin are a definite plus, as is the additional equipment they usually carry when compared with a single. And for people who are more comfortable with two engines, they are great. But whatever value might be found in having two engines is there only if the pilot is very sharp. A constant refresher program is necessary. So is an awareness that the fact that the airplane is a little larger and more powerful doesn't offer any measure of protection against pilot error.

Setting the stage is one of the most important parts of instrument flying. Think about the big picture. Study the real risks. Recognize that instrument flying is a challenging mental exercise. Carefully evaluate every little glitch in your flying. And remember, a big part of staying IFR refreshed is related more to thinking about the right thing at the time than to anything else.

CHAPTER 2

Preparing for an IFR Flight

FIRST PART   Patrick E. Bradley

As a VFR pilot, I always thought it exciting and romantic to decide on the spur of the moment (and on a clear day) that I would fly to some distant place for lunch, or for the afternoon. I would draw a line on my sectional, call weather to verify that the sun was shining (and would continue to do so), check the airworthiness of my airplane (not worrying too much about instruments, since I most probably would not depend on them for much), and fly off. I knew that if the weather unexpectedly deteriorated, I would terminate the flight. I knew that if I was reasonably vigilant, I could follow landmarks to my destination and back. They were simple flights, and I knew that if the unexpected happened, I would just spend the afternoon, or part of it, on the ground, making whatever new plans were necessary. I wasn't determined to reach my destination, I just wanted a reason to fly.

Shortly after I began instrument training, I learned that "seat of the pants" thinking did not lend itself to safe flight through the clouds. My instructors constantly emphasized the importance of methodical preparation. No stone should be left unturned where it could jeopardize the safe completion of a flight. There were regulations, procedures, and forms to help make sure that I knew where I was going, when I would get there, how much fuel I would consume getting there, the weather I would encounter, and where I would go if I could not reach my destination. My instructors taught me steps to eliminate variables systematically before I ever started the engine. Midway through an instrument flight was no time to start worrying about whether there would be enough fuel, and it was no time to realize that your glideslope indicator was on the fritz.

So I gave up the romantic notion of grabbing a chart and flying through the clouds to some unknown area on the spur of the moment. I dedicated hours to preparing for my flights. I calculated, for any given point along the route, how much fuel I had used and had remaining; I listed the frequency for every navaid. No blank on my flight planning sheet was left unfilled. And do you know what? I was still unprepared — not completely, by any stretch of the imagination, but there still were times when I was caught needlessly unawares. I felt a little betrayed. I thought that if I filled in all the blanks, then I should know everything I had to know about a flight, but I was wrong.

What happened during these early flights, and more of my current flights than I would like to admit, is that I fell prey to a sort of robotism. Like a human calculator, I would fill in spaces thinking that an IFR flight must necessarily run according to the rules set by forecasters and my little forms. I still use those forms, because they are valuable guides. But now I try to look at them as a general, and not necessarily infallible, tool for helping me deal with the unexpected.

When I began to notice that much of my flight planning time was wasted time, I fell prey to another evil: cutting corners. I figured that if the winds are almost always going to be different than forecast, then why not just forgo calculating speed altogether. If I used some reasonable speed, then I would be close in my approximation of when I'd reach my destination. Another corner I frequently cut was in planning my route of flight. I would dash down the airways that looked fairly reasonable and then wait for my real route to be generated by a computer. When I labored over the route, trying to stay away from mountains or to find the shortest path only to find that ATC had other plans, I couldn't see much point in really paying particular attention to the airways I had filed. I would take what I was given and leave it at that.

My philosophy during this phase came down to doing what was necessary to get off the ground safely, and to dealing with changes as they came. I saw experienced professional pilots doing this frequently (or so I thought), and saw no reason why the same technique wouldn't work as well for me. Cutting corners saved a great deal of time that I generally needed to prepare for other elements of the trip, such as arranging for a friend to pick me up at the airport, or for a hotel at the destination.

Since then, I've realized that neither system was really satisfactory. When I cut corners, I saved lots of time, but more often than not, I was completely unfamiliar with where I was going on a first-time flight. When the amended clearances came, I didn't have the benefit of having looked over potential routes closely enough to know where I *could* be sent. In fact, I was once cleared en route to a navaid that I not only had never heard of, but couldn't locate. I put myself in the position of either suffering the ultimate embarrassment of admitting my plight, or struggling to find the VOR as I flew aimlessly. I asked for a heading. That was the point at which I decided I'd really have to come up with a better system for getting myself set for a flight — especially into unfamiliar areas.

First, I decided that even if I couldn't predict my route with certainty I would try to get a general idea of all the options. I like to check the lay of the land in areas with which I am not familiar. I take in the names of VORs on the route and surrounding the route. If one possible route is significantly longer than another, I make a mental note of it. Another technique I've begun to use is to keep a record of past clearances. If I've flown a particular route once before but can't remember it off the top of my head, I look it up. It almost always works.

Second, although I keep a log of distances for my proposed route, I don't go through the chore of calculating estimated time or speeds for each leg of the route. I take a target speed based on an average of the forecast winds, and work from there. If I know that fuel won't be a factor on a particular flight, I might not calculate my time and speed for each leg of the flight. If fuel is going to be a factor, I make time and speed calculations a priority. I know then that if I go slower than my target speed, I will have to stop short of my destination. If I am faster than my target speed, then I will arrive at the destination with more than my hour reserve.

One area, though, where I have never cut corners and know of no better system than going strictly by the book is in familiarizing myself with the airplane I will be flying. I rent airplanes and often fly different types. It is something I consider a real disadvantage in planning a flight. To fly a number of different airplanes safely, I always get hold of a pilot's operating handbook before I get checked out. After I am checked out I adhere pretty strictly to the handbook for all preflight planning.

I know that fuel consumption for a particular airplane can vary from the handbook, so I make sure that I never violate my hour reserve when planning an IFR flight. For planning, though, I go by book fuel consumption for the particular conditions in which I will be flying. Cruise speeds, weight and balance calculations, and all other calculations also come right from the book. Once I have flown an airplane for a while, I might know that I won't make book speed and will then make my own changes, but I have seldom flown an airplane in enough different conditions to be able to do that confidently. Other facts such as usable fuel and fuel consumption rates are easy to confuse from airplane to airplane. It always is important to be aware of the V speeds for different airplanes. I've found that just because I knew them when I was checked out doesn't mean that I'll know them when I return to the airplane several weeks or several months later. And finally, there are emergency procedures. I've been amazed at the divergence in procedures for different types, and I am very bad at guessing best glide speeds for different airplanes — a definite memory item.

I sometimes think that I go a bit overboard with the operator's manual. But above all, I need the psychological support of knowing my airplane's specs. I also won't immediately fly an unfamiliar airplane into instrument conditions if I can help it. When I have been flying a particular airplane, I feel that I am more familiar with its quirks, leaving myself open to fewer chances for surprises. If a com works intermittently, at least I'll be mentally prepared for it to cut out for a while. Checking a rental airplane's squawks is no assurance that they are either all fixed or that they are the only malfunctions you will encounter. I will generally fly a slower or less comfortable airplane if it is one with which I have had some experience.

Preparing for an IFR flight, for me, consists of preparing myself for the variations that can and probably will come up in the course of a flight. When I spent all my time mechanically filling in spaces on a flight planning sheet, I worked on the assumption that an IFR flight is a static entity. I believed it would proceed as I had planned simply because I had planned it to go that way. When I found that this nearly never worked, I erred to the other extreme. I gave up trying to figure out the most likely changes and how I could prepare myself for them. I left my planning entirely to the cockpit.

Now, although I know I can never try to prepare myself for every eventuality, I try to guess where changes are most likely to occur and can often predict them ahead of time. I get myself mentally psyched to deal with all obstacles during the flight, and this really begins on the ground. When I start the flight unsure of the airplane I'm flying, the route that I'll be taking, or the airport that I'll be flying into, it's difficult to really respond authoritatively to new wrinkles. Of course, this doesn't mean gleaning every iota of information available on a particular flight. That could take as long or longer than getting into your car and driving to your destination. It means attempting to focus on pivotal areas of concern: In poor weather, this focus might be on determining where the weather is better; in a new airplane, it might be on proper operating procedures; on a new route, it might be on terrain or navaids or the destination airport. Sometimes it might be all of these things, other times it might be none of them. Determining the essential areas of concern, I think, is the number one element of good flight planning.

SECOND PART   Richard L. Collins

There's no question that going at an IFR flight without proper preparation can lead to difficult situations. A constant review of the way planning is done, along with a good critique of planning for each flight, is the way to stay on top.

An example of how a lack of planning creates confusion came to me at Regina, Saskatchewan, one summer morning. I had left my airplane there a few days before because one magneto was dead and a replacement wasn't to be found quickly in Canada. I reluctantly went home on the airline and returned with a mag. The flight from Regina to the U.S. had been planned before, and the day I returned to pick up the airplane I made some impromptu changes in the planning while waiting for the technicians (engineers in Canada) to change the ailing magneto. An IFR flight plan was filed, and when they finished I did a careful preflight and hopped aboard.

It was at this point that I should have paused and reviewed the plan. I filed direct to Minot, North Dakota, 210 nautical miles away. That's easy. Just point it in the right direction. I remembered calculating the bearing from Regina to Minot before landing at the former, and the number was in mind. I didn't recheck it. After takeoff on Runway 12 at Regina I asked the tower for a right turn on course. They approved, and I started a right turn to 210 degrees — the number I had set on the heading bug.

Even an elementary knowledge of geography strongly suggests that this was the wrong number. I had transposed numbers in my mind — the correct heading was 120 — and didn't catch it until looking at the horizontal situation indicator. Then it became clear to me that if I was going home to the east coast there was no way that the first leg would be flown on a southwesterly heading. Needless to say, I felt quite dumb. A more careful review of the plan before takeoff would have prevented the error.

ELEMENTS

Weather is a big thing in IFR flying, and even though what we experience once aloft is what we get — regardless of forecasts — the thought given to weather before flying can make planning much easier and can give clues that avoid surprises.

A continuing curiosity is a basis for flying-weather deliberations. Watch the weather on TV every day or look at a weather map in the paper, and you have a continuing idea of how weather systems move across the country. Add to this a sampling of what is going on outdoors for the effect the patterns are having on the local weather. The basics are then in place.

The local surface wind flow tells a tale of weather. The meaning of wind varies with a location's relationship to mountains and bodies of water. We thus develop a better ability to judge weather by surface wind at home and frequent ports of call. For the places I fly most (east of the Rockies), considering local wind direction starting clockwise with north, the weather deteriorates through a wind from the south, where it starts to improve. Improvement continues back through north. Anything between south and west carries with it the chance of a squall line in the appropriate season, because wind from that direction is the precursor to a cold front. When I'm at home in New Jersey, a freshening breeze from an easterly direction hints of low pressure formation along the coast — always an interesting development.

Running outside, wetting your finger, and determining wind direction doesn't combine with a look at the TV weather to make you a weather wizard. But the pilot who does maintain a continuing interest in weather is likely to have the easiest time planning a flight and getting a meaningful weather briefing — because he's constantly thinking about the big picture.

BACK TO THE CHARTS

Foreknowledge of weather is a big part of the flight planning process. For starters, unless you know something about the weather, the requirement for an alternate airport can't be approached intelligently — nor can the question of the amount of fuel required for the flight.

I learned a lot about weather and flight planning during the time the general aviation reservation system was in effect. To be assured of an IFR reservation, you had to file twenty-four hours in advance. No terminal forecast extends that far into the future, nor do the readily available winds-aloft forecasts offer such long-term predictions. There were two choices: Flights could be planned without regard to weather, or I could make my own weather projections and plan on that basis. Because many trips involved a fuel stop and thus two reservations, and because something is better than nothing, I saw no choice other than to make my own long-term forecasts.

We really need know only a few things to anticipate weather that might have a profound effect on a flight.

Is there any likelihood of a storm system — one that would result in heavy snow or widespread severe thunderstorm activity — affecting the route? I find good clues to this on TV. If the people using weather forecasting computers suspect that something big is brewing, they like to say so as far in advance as possible. And they like to say it to the largest possible audience. More often than not, storms don't develop precisely as projected on the morning TV shows of the day before, but that doesn't matter. The word that it is possible arouses our thinking about contingency plans. When planning twenty-four hours in advance, I'd go for shorter legs and two fuel stops instead of one to allow for diversions that might be dictated by a predicted storm. Also, *storm* and *headwind* can be synonymous, especially if you are headed in a south-westerly direction.

There are three notable ceiling and visibility conditions that play on IFR planning. The first is 2,000 feet and 3 miles, the forecast condition at a destination below which we have to have an alternate airport. The second is 600 and 2, generally the alternate minimums for an ILS-equipped airport as well as a benchmark that makes questionable a nonprecision approach if that is all that's available at the destination. The third is 200 and ½, usually the minimum conditions for successful (and legal) completion of a normal ILS approach.

PRECIP

If, twenty-four hours ahead, it looked to me as if there would be precipitation at any destination, I'd then consider that an alternate was necessary. There is not likely to be a forecast of 2,000 and 3 or better with precip also in the forecast. Looking for an alternate twenty-four hours ahead, I'd project surface map features and guess which areas would be south of any warm front, or west of a low or cold front. I've never come up with an intelligent way to guess at below 200 and ½ conditions from twenty-four hours in advance. My only rule of thumb was for arrivals in areas of bad weather in the period of a couple of hours after sunrise and after sunset. The rule simply said "beware." Fly with a lot of fuel.

Advance weather planning, whether started twenty-four hours or two hours ahead, is necessary if one is to be able to do things in any logical sequence and to minimize work load toward the end of the planning process. You simply must have a weather plan *before* calling the flight service station or accessing aviation weather from any source. Without a plan, you don't know what to ask for. I've found that it works best to make the plan based on general knowledge — there's little difference between general weather and aviation weather — and then go for the details on the basis that you are going to make the plan work. If something big crops up in aviation forecasts that you didn't expect, then it's back to square one. Start over. Usually, though, the TV briefing gives the big picture as well as any aviation source does.

The actual planning of a flight can take many forms. Some people like detailed forms, some don't. Whatever turns you on. The important items are the distances and the routing. Some pilots carefully take all that information from the chart and meticulously record it on a form. Other pilots just use the chart. Before deciding one system is better than another, consider the benefits of each.

A detailed flight log often includes the distance left to go from any given point along the route. If the flight is anywhere near maximum range, this a critical item. Current ground-speed and distance to go gives a projection of how much will be left in the tanks on landing. You can add the numbers en route, but perhaps it is best done in advance. I've never had a form that incorporates "distance remaining," but I've always thought that perhaps I should. What I use instead is a minimum acceptable groundspeed. Calculated in the planning phase, this is the groundspeed en route that is necessary to get the airplane to the destination with the required fuel reserve. Anything below that means a new plan. No kidding. One knot below requires a new plan. Often I do calculate two minimum groundspeeds in advance — one at normal-cruise fuel flow and one at long-range cruise. Then if normal cruise isn't hacking it I go to long-range cruise. Failure to meet the goal there is a mandate for further change in plan.

ROUTE

When selecting a route, we can get some clues from charts on what might be preferred by air traffic control, but there's no guarantee. Certainly if there is a standard instrument departure and a standard terminal arrival route, connecting the two with the most direct route will probably result in a flight plan bearing some resemblance to the actual clearance. Lacking those, a logical departure coupled with the most direct route to an initial approach fix at the destination is the best we can do. If there is a published preferred route along the way, it might be used. But the ATC folks don't seem to put much stock in the preferred route; the real keys are in departing over specified fixes and arriving over specified fixes. Trouble is, these aren't published. For example, if you are coming toward Philadelphia from the west or southwest you might as well file over Harrisburg because you'll eventually get a clearance to do just that regardless of the initial clearance issued. But this isn't on any chart. That is one reason not to be bound to a plan. Things will change and you must be flexible. Unless you fly in the same area all the time and know from experience the preferences of ATC, the flight you plan and the clearance you fly may be different.

TERRAIN

We've put an overview of weather onto the route we are going to fly, but there's still something left to do. It's good practice mentally to position terrain beneath the route. If you are flying from Minot, North Dakota, to Grand Forks, this might not seem important. But the accident reports are full of the remains of airplanes the pilots of which did not have good knowledge of the terrain and descended below a safe altitude only to find a fog-shrouded hillside. An unfamiliarity with charts can contribute to this. Part of any refresher is a session with charts, to make sure you understand what everything means. Equally important is appreciating why some things are as they are.

Ridges near airports have over the years taken quite a toll of airplanes. The most famous case was that of TWA 514, which hit a ridge west of Dulles airport while on an instrument approach. The controller cleared the crew for the approach, there was some discussion on the flight deck of what the chart meant, and then the airplane descended to an altitude below the top of the ridge that it hit. It was not flying on a published route where the safety margin is carefully calculated. The message that has been repeated many times since then is related to the importance of safe altitudes and published routes. A companion message, not so often repeated, is that a knowledge of the terrain can keep you from doing something illogical even if you misread the chart.

In addition to arrival accidents, there have been cases of airplanes departing and flying into mountains. Again, if a careful mental review of terrain had been part of preflight planning, these accidents would not have happened.

The importance of understanding what the numbers on the chart mean was underscored in a FlightSafety International refresher ground school. We happened to be looking at the Jeppesen approach plate for Tucson International Airport. In question were the numbers in parentheses below the minimum altitude for the flight segment. The class was quizzed on what the numbers meant; many thought they meant the minimum terrain clearance altitude for that segment. They actually are the height in feet above the touchdown zone elevation at the airport. In the case of Tucson, if a pilot took that to mean a terrain clearance altitude he might well rearrange some of the rocks on one of the mountains in the area. A pilot who had visualized terrain in relation to the route flown wouldn't make that mistake.

THE AIRPLANE

With general weather and terrain considered, what else is there to planning before getting on with the final briefing and the launch? We might ask ourselves if we really understand the airplane to be flown. Professional pilots have, in initial and refresher training, the benefit of ground school on their particular machine. We don't have ground school to cover the nuances and the things that happen once in a blue moon. All we have is the pilot's operating handbook — which may be excellent or which may have been written by a legal department that has little experience with airplanes in actual use. Regardless of any perceived shortcomings, the POH contains a wealth of information. And in IFR flying, where we are depending a bit more on the airplane than on a clear day, it is foolish to fly without a complete understanding. From the best glide speed for maximum range (which could come in handy if gliding through an instrument approach after a power failure) to the emergency landing gear extension procedure (which would always be disorienting in the clouds), there is a lot there to learn. I've always thought there should be a quiz in the back of every POH to make sure the pilot has studied it and understands some of the fine points of the airplane. Here's what I would put in the quiz on a P210, the airplane that I've flown more hours than any other. I'll give the reasons for each item. You can develop a quiz for your airplane.

1. If the induction air filter is clogged and the alternate air door opens, what is the maximum manifold pressure that might be lost at cruise power? (The answer is a whopping ten inches. I had flown the airplane 2,350 hours before actually experiencing this — which just goes to show that you learn something new every day.)

2. If the vacuum pump fails, what instruments and functions are lost? (All deice boot operation, vacuum instruments, and the heading-hold function of the Cessna 300 autopilot. If I had a more sophisticated autopilot, the whole thing would go after a vacuum failure.)

3. If the low voltage light comes on, likely indicating a failure in the charging system, what should I do first? Second? (First, turn the battery and alternator switches off and then back on to see if that will bring the alternator back on line. If not, turn the alternator switch off, turn nonessential radio and electrical equipment off, and land as soon as possible.)

4. If gliding, power off, after an engine failure, at what altitude would I want to cross an outer locator 5 miles from the runway of intended landing? At what altitude would I want to fly over a runway on an overhead approach to a normal power-off pattern. What speed should I fly? (The airplane will glide approximately 1½ miles for each 1,000 feet of altitude. So, you'd need about 3,333 feet to glide 5 miles. Because the descent angle can be increased with landing gear and flaps, but not decreased, and because headwind would cut into the distance, I'd want to stay a bit on the high side. I'd leave the marker at 4,000 feet unless there was a very strong headwind. Then I'd

add altitude. On an overhead approach, I'd figure about 3½ miles of flying back around to the runway, plus or minus a little. That would mean 2,333 feet. Add to that some for the turns and the fact that you could work with the approach slope based on visual observations and I'd start at 3,500 feet. The best glide speed for an average weight is 83 knots, gear and flaps up.)

5. If the static source is blocked and I have to revert to the alternate source, what allowances must be made? (Approach 7 knots indicated airspeed faster than normal and add 50 feet to the MDA or DH.)

6. Engines don't run well on water, VFR or IFR; how many quick drains are there in the fuel system? (The answer is five. If you fly a 210 and think there are only three — as do probably 80 percent of the 210 pilots — check again.)

7. At what point does weight become critical? Balance? (Pilots seldom calculate weight and balance for every flight, relying instead on guidelines. Four persons and baggage are fine in my P210 unless all four are linebackers. The balance is okay at that so long as passengers use the middle instead of rear seats. Anything farther aft or in excess calls for a detailed weight and balance computation.)

These are some samples. The answers are in the POH — usually. Some things we have to learn about our airplanes through experience. The primary thing is to have an active curiosity.

INTERFACE

When flying a more complex airplane, there's some interface between the operation of the airplane and flying IFR that calls for advance planning. After watching some relatively new pilots wrestle with the combination of a new instrument rating and a complex airplane, I worked on a system to help them deal with both at the same time. Certainly it is a mistake to think that flying IFR in a turbocharged single is the same as flying IFR in a simple single.

The primary problem I saw was remembering to do the things that the airplane needed while at the same time accommodating the demands of the IFR system. A primary example of the need was illustrated on an average departure from my home airport (Mercer County, in New Jersey) and a return to that airport.

Leaving, the climb is always made with a lot of restrictions. It's common to level off at 3,000, 4,000, 6,000, 7,000, and 10,000 before being cleared to a cruising level in the high teens. That means the throttle, prop, mixture, and cowl flaps have to be dealt with twice for each altitude — once when leveling and once when resuming the climb. With new pilots it was a time of confusion until some advance planning was applied. In effect, you have to think in terms of a "level off" and a "resume climb" checklist. The same is true on arrival. The descent is usually in stairsteps, and there are certain things that must be done each time.

This led me to have pilots draw a profile of a flight before flying — including the stairsteps on the climb. After takeoff, write in the items that are involved in the transition to cruise climb. Then throw in a couple of level-offs with the appropriate checklist: level off — pitch, throttle, prop, cowl flaps, mixture; resume climb — cowl flaps, pitch, mixture, prop, throttle. Draw it all out and write it down. Cover every step. Plan every aspect of the flight, and it's less likely that you'll omit something or follow an incorrect sequence.

HARDWARE

Other airplane-related planning should be directed at the particular requirements of IFR in the specific airplane.

We've learned that systems failures cause problems in single-engine IFR, for example, and this needs to be thought through *before* flight. If the weather is relatively good and the flight is a short one with a lot of options (such as flat terrain beneath and a number of airports in the area), then an alternator failure might not be so critical. It would be possible to turn off all nonessential gear and have enough juice in the battery to scurry to the nearest airport and land. Switch to a cruddy-weather, long-haul flight in more sparsely occupied territory, though, and a lot more time might be involved between the failure and a landing. That's when a handheld, independent battery-operated transceiver might be among the most valuable things in the airplane. The planning phase is where things like this are best turned over in mind. That way, the necessary gear won't be at home (or on the dealer's shelf) when it should be in the airplane.

MEET ACE PILOT

IFR imposes unique demands on pilots as well as airplanes, and this is where the word *refresher* has greatest meaning.

Can the pilot handle everything that might come up on the flight being planned? Is the pilot okay for normal operations but rusty on everything else? Certainly nobody goes through the list of all possibilities before every flight and checks off pilot qualifications for each. There are simply too many items to consider. But we can go a long way toward anticipating bad stuff by looking at the items that have proven to cause the most trouble.

It is confidence in the ability to handle an IFR flight after some things go awry that helps psyche a pilot up for an instrument flight. And, getting psyched up for the flight is important because punching into clouds is not something you want to do with nagging doubts in mind.

We'll cover the handling of emergencies in detail in a separate chapter. Here, in the preflight area, I'd just like to explore one thing to illustrate how we can help ourselves by relating our equipment to our ability and to the conditions that are expected to exist during the flight.

Vacuum pumps, or rather the failures of vacuum pumps, have been a big subject recently. And there have been accidents resulting from pilot inability to fly with the instruments still in service after a pump fails.

I had an airplane that had a rash of pump failures; they came often enough to illustrate over and over to me the things that a pilot can be called on to do after the failure of a vacuum pump.

To begin, as best I recollect, I had three vacuum pump failures in 11,000 hours of flying before I got the airplane with failure-prone pumps, a P210. One failure, my first, was in a 250 Comanche. I was IFR. It was a hazy day, and I got a bad case of the leans before finally settling in with the turn and bank. I canceled IFR, descended to a relatively low altitude, and continued on home. The second failure was in a Cherokee Six, in good weather. The third was in a Turbo 210 (not mine), and it was probably the most impressive of the three. The airplane was one of Cessna's avionics demonstrators, with their best autopilot and a flight director. I was in good weather when the pump failed but had flown it in a lot of weather on the previous flight. The failure meant a transition from a full autopilot and flight-director system back to the most basic equipment — a turn coordinator in this case, along with airspeed, altimeter, and vertical speed. How would that have worked in bumpy cloud?

When I got my P210 I practiced some partial-panel flying in it, in case I had a vacuum failure. Was I ready? I certainly didn't worry much over it. I learned to fly instruments on partial panel and always thought that I could do a passable job.

The first failure didn't cause me to give the subject a lot of thought. I had only to descend through a thin layer of clouds, and that was no problem. A subsequent failure left me with some instrument flying to do, and a later one left me with a lot of bumpy instrument flying without the benefit of a full panel. It was at this time that I balanced the airplane's capabilities and mine against the weather that I fly in, and decided that good planning dictated some additions to the panel. I added an electric directional gyro and later also added an electric artificial horizon for complete redundancy in instrumentation. My partial-panel ace status simply didn't extend to an airplane that flies so high, that is relatively complex to operate, and that I use in all kinds of weather. Were I flying a Skyhawk or Warrior, I'd be perfectly happy with partial panel plus a little skill as a backup system for vacuum. In either airplane, giving advance thought to the question would help me fly with confidence in my ability to handle the challenge. The worst thing we can do is fly with the thought that a failure of any system, or part of the airplane (short of primary structure), would automatically mean big trouble.

TAKING STOCK

Most pilots are able to reduce preflight planning almost entirely to a yes or no, black and white process: *The airplane is fit and I understand the airplane. I am fit, proficient, and ready to handle the flight and am not depending on any failure-prone system or piece of equipment for the safe completion of the flight. The route has been carefully analyzed and the flight plan filed.* Everything is there. Almost. If there is any nagging doubt that affects a large percentage of pilots, it is in relation to the weather.

Every pilot has moments of pause about weather. This is natural. The pilots who have real trouble dealing with weather are ones who entertain an unaffordable luxury — the thought that there is some mystery about weather. This is a pure cop-out, an admission that the ambition to learn about weather simply isn't there.

It's not difficult to be weather-wise. All it takes is an understanding of the basics, along with the common sense to accept and deal with the weather as it exists without regard to forecasts, hunches, or guesses.

Forecasts   Forecasts are good items of information but should be taken as that and that alone. The fact that you have a forecast of good (or bad) conditions is no guarantee of anything. It is but a computerized product that projects an expectation based on a number of factors. Some of the known quantities fed to the computer are very old. The forecasts are of necessity very broad, covering a wide variety of things that may or may not happen. Perhaps some of the broadness of forecasts is for legal reasons: The forecasters would naturally want to cover every eventuality. Many times, forecasts are incorrect because the information which they are based on is incorrect. The upper wind patterns that steer and influence the development of surface systems are often not as the computer anticipates. When the upper-wind forecast is wrong, everything else tends to follow. There can be times when forecasts over a wide area are incorrect, and the system is slow to catch on.

The way a pilot deals with incorrect forecasts is simple: Use them as what they are — estimates made by a computer that has no window to look out. If there are no thunderstorms in the forecast but there's static on the AM radio as you drive to the airport, believe the AM radio. If it's nighttime and no fog was forecast but there are halos around all the streetlights, believe the halos, not the forecasts. Often, perhaps more often than not, the weather is better than the forecast. Then proceed based on what you see — but proceed with a wary weather eye.

The fallibility of forecasts was brought home to me one summer day when I was en route to Florida. My son checked the weather for the flight; he said that the FSS person said things would be okay — some scattered thundershowers along the coast was about all he had on his list. The wind would be pretty good, and the forecast for the destination was excellent.

There was some crashing on the AM radio as we drove to the airport, but the optimistic, standard-broadcast station forecaster was sticking to his guns for a nice day (it was Sunday), saying that there were a few showers to the north but that clear skies should be along after a while. He had the same information as the FSS person.

The weather at the airport looked okay — hazy and hot. The sun was out. We called for a clearance; it was along a different airway than filed. Then there was a delay getting a departure release. The explanation for the delay was that the airway was saturated with traffic. I suggested that they had routed me down a different airway than filed and if they would let me use the airway I wanted, maybe it wouldn't be saturated. There was some discussion, and then our departure release came through. We were soon to learn that the airway we had originally filed for was indeed saturated — with severe thunderstorms. The reason for the delay was that all aircraft were being routed around on the other airway, and that was creating delays. To boot, there were a lot of thunderstorms along that route, and all aircraft were deviating a great deal. Additionally, we encountered moderate icing at 16,000 feet and had to descend. The wind-aloft forecast was incorrect (it was out of the southwest instead of the northwest as forecast), and this went hand-in-hand with the inaccuracy of the other forecasts.

So, on a flight that was a piece of cake in the planning stages, I found the airborne weather radar and Stormscope in the airplane very handy. The deicing equipment was put to good use as well. Back in New Jersey, what had been forecast as a clear to partly cloudy day with a chance of showers ended with a total of over two inches of rain. The forecasts were quickly replaced by what was actually happening. If that happened once in my life, it would be one thing. But it happens frequently enough to make me basically distrustful of all forecasts. And I think that this distrust is a healthy thing. It has kept me from being surprised at below-minimum conditions when the forecast was good, and it has kept me going when the forecast was terrible but the weather looked good.

Hunches   When we question forecasts, it has to be on the basis of something better than a hunch. It has to be on the basis of something that is observed to be true: fact. If the weather is forecast cruddy but the ceiling and visibility are okay, the radar does not show the expected precipitation, and the wind is from a favorable weather direction (usually westerly), then maybe it is okay to forget that forecast of inclement conditions. But all the good factors have to be there. If, for example, the ceiling and visibility are okay for you for IFR but the wind is still out of a southerly direction and the radar chart shows a lot of activity to the west, you would want to carefully consider alternatives — especially in an airplane without weather avoidance gear such as radar or a Stormscope.

Guesses   These fall in much the category of hunches but with an important difference. I think guesses come when a pilot is operating in a vacuum of no information. "I guess I will be on top at 6,000 and can see the storms visually." That can be a very uncomfortable guess. Usually when there is convective activity, it's tough to plan on being on top and seeing all the weather.

TAKEOFF ALTERNATE

Another thing that helps make the initial planning phase less critical, at least in feeling, is planning a takeoff alternate. In operating multiengine airplanes, takeoff alternates relate more to engine failures. Where would you go and land if an engine failed soon after takeoff? In a single, the engine-out part is a rather meaningless exercise, but in any airplane a takeoff alternate is worth having in case something beside the engine makes you want to land. The alternate might be the airport of departure, or it might be a nearby airport, or it might be an airport in the direction opposite that of the planned flight.

One thing the takeoff alternate does is give a place to go if you depart and then find the weather not as forecast and not to your taste. For example, if the forecast led you to expect a smooth flight in or above stratus clouds but your initial cruising level is wet and bumpy — indicating that the clouds are cumulus instead of stratus with their tops likely well above your cruising level — then you might not want to continue. That's when a takeoff alternate would come into play. Another use is to cover oddities other than weather — a gas or oil cap left off, a belt out the door and beating on the fuselage, a chart left at home, a strange noise in the airplane, or a passenger who wants to go home.

Back to weather, the freedom from takeoff minimums is something that we want to maintain in not-for-hire operations. Part of keeping this freedom is responsible use; always having a takeoff alternate is a key to this. If we make a 100 and a quarter takeoff, have a problem, need to land as quickly as possible, and create a scene because of a lack of both a plan and knowledge of conditions in and around the area, that makes a mockery of a responsible operation.

One final weather question should be considered. Much is made of the go/no-go decision, but I have often suggested that this should really be the start/continue decision. Is it okay to start? Then, is it okay to continue as planned? I have found time and again that it might appear fine to start a flight when there is some question about weather farther along the line. But I have also found that I can make pretty good decisions as I go along. There is usually some way to work around big en route problems. The "continue" answer might not be favorable straight ahead so I detour a lot — perhaps more than some other pilots. I like smooth and uneventful rides. A few or a lot of extra miles matter not. And it's often possible to deal with some really hairy weather without straying all that far from the planned route. Both weather and airplanes are dynamic. There can be a lot of rearranging of both as you go along. Just think of the bad things en route — generally ice and thunderstorms — as solid objects in the sky, and always plan to avoid them.

PREFLIGHT BUSY WORK

The proper preflight of an airplane is well covered in other places so I won't dwell on it here. Just let it be said that it's doubly important to *know* that everything is in order *before* punching into a cloud.

There are things other than tire kicking and oil checking to do before an IFR launch that will make life easier. It's a procedural exercise, flown according to directions.

Plans were made when figuring the route, but they are worth reviewing in the airplane. Get the charts arranged and look at the first part of the route. Visualize what you'll be doing in the initial phase of the flight. An example: "I'll probably depart on Runway 24. It'll then be a left turn to a heading of 180 to intercept the 150 radial of Yardley, which is the first part of the standard instrument departure. The usual initial altitude assignment is 3,000 and the departure frequency, 123.8." If you have old-fashioned radios you can preset frequencies at this time; with the newer electronic readouts you have to wait until power is on the equipment. If the clearance comes as expected, fine. If it doesn't you haven't lost anything.

Here I'm going to make yet another sales pitch for a piece of equipment and give you a good use for it. All pilots flying IFR in airplanes with single electrical systems should consider having aboard an independent battery-operated transceiver. There are many different ones available at a cost reasonable to an aircraft owner, or FBOs could make them available for rental IFR aircraft. If you have one of these, you can listen to the ATIS and get your clearance *before* starting the engine. You would be surprised how much easier it is to digest information and copy a clearance without the engine running. It's a calmer time. Also, if the clearance is different from the filed route, this gives you a moment to study the new route in a relaxed atmosphere. It also serves as a test of the standby radio.

There is nothing unusual about getting a clearance before engine start, either. The airlines do it all the time. And after you get the clearance, there's no big rush to depart. The IFR departure from an airport is okayed in two stages. The clearance means that this will likely be your route, at least for the first part of the flight. All involved controllers have been given the information. After you get the clearance you still have to get a departure release — meaning that the system is ready for you to actually fly. They don't start working on the departure release until they know that you are ready to go.

Finally, before going, survey the air into which you will be flying and review the departure in relation to the terrain in the area.

There are places where an incorrect action soon after departure can result in a terrain-related disaster. A good example is found in a business-jet crash. Let's review it before going on to the next phase of our IFR refresher.

The flight was cleared to take off on the Palm Springs Runway 30. The ceiling was 3,000 feet, and, as any pilot who has flown out of Palm Springs knows, there are mountains, up to an elevation of over 11,000 feet, northwest of the airport, about 20 miles away. The IFR departure procedures acknowledge this by calling for a turn to an easterly direction soon after takeoff. Any pilot visualizing the area before takeoff would give a lot of thought to those mountains.

The crew got a rather confusing clearance that they incorrectly interpreted to mean "maintain nine straight ahead."

A bit later, the crew asked the controller the following question: "We're maintaining nine on a heading of 310, what's our clearance from here?" The controller told the crew to expect further clearance crossing the 20-mile fix, which really wasn't applicable because they were not tracking any radial at that time. The controller later testified that he thought he heard the crew report on a heading of 030. (It was a nonradar environment so the controller had no way other than the pilot's report of knowing for sure the heading of the aircraft.)

The airplane flew into high terrain at 9,000 feet, as would any airplane flying runway heading off Runway 30 out of Palm Springs.

The National Transportation Safety Board found the probable cause to be a misinterpretation of a clearance and subsequent instructions on the part of the flightcrew. One Board member added that the accident occurred, in his opinion, because of "nonstandard and nonprofessional clearances and transmissions on the part of both the controller and the pilot. If the proper language had been used, the accident most likely would not have occurred."

All that might be true. But, again, the whole thing could have been prevented by a little analysis of the clearance versus the terrain. There was a big mountain straight ahead. It was obscured by cloud but it was (and is) still there. Runway heading was an impossible way to go. A pilot's perception of rock-filled clouds can often save the day.

CHAPTER 3

Getting into
the IFR System

FIRST PART   Patrick E. Bradley

After my instructor told me that I was qualified to fly IFR, and after the FAA told me that I was qualified to fly IFR, I had to struggle with the question of whether *I thought* that I was qualified to fly IFR. Whether conditions are within my level of expertise and my airplane's capabilities was and is a decision with which I must grapple before embarking on any instrument flight. And it may well be one of the most difficult decisions a new instrument pilot has to make.

At first I thought that making the decision to enter the IFR system simply involved analyzing the weather, my own level of skill, and the capability of my airplane in as objective a manner as possible. In some cases, it would involve swallowing my pride and waiting for better conditions. In other cases, it would involve summoning up the courage and the self-confidence to take the leap. But I found that when I made my decision this way, I seldom flew IFR except in VFR conditions. I couldn't analyze my skills objectively or subjectively when I had never flown IFR alone before. And after I had flown IFR alone several times, I still couldn't tell whether the forecasts were for conditions too tough to attempt, or whether briefers were overstating the severity of the situation. More than once I canceled a flight on the basis of forecast conditions at my destination. Later I would discover that the weather actually had been good enough to have completed the flight. Few things are as frustrating as that scenario.

So I languished, stalled at the threshold of an IFR flight. I began to have serious doubts as to whether all the time and expense of getting that wonderful IFR rating was worth it. I felt that I was no nearer mastering adverse weather than the day after I got my private. The final straw came one day in April. I had planned to fly from Teterboro, New Jersey, to Nashville, Tennessee, for a business meeting. It wasn't the type of meeting that could be rescheduled easily, and I prayed and prayed and prayed for good weather. Of course, on the day I was to leave, the weather was miserable.

A call to flight service confirmed the obvious, and the friendly briefer assured me that the weather was as wretched down south as it was up north. I put off a final decision until I got to the airport. With any luck, the clouds might part. It wouldn't be the first time that the forecast was wrong.

I got to the airport before the FBO opened, and the Archer II that I had scheduled looked puny against the dark clouds. Standing in the drizzle I mulled over the question of whether I should really go through with the trip. Then the FBO manager arrived and opened up the building. I called up flight service again and got more of the same. I couldn't pin them down to anything really horrible except that I could expect icing at about 7,000 feet. Icing conditions are horrible enough, but there was a good 6,000 feet below the freezing level in which I could fly safely. I mulled it over while the line boy filled up the airplane.

Finally the balancing and the weighing and the vacillating got to be too much for me. I decided that I would at least start this flight. No thunderstorms were forecast, I could avoid the ice, and if things got too bad, I could quit with the satisfaction that at least I had tried. Try I did, and I reached my destination — but not without some second thoughts. Conditions en route were significantly worse than forecast, but at that point I was in the system, and I was keeping my head well above water. I made a lengthy detour around a thunderstorm with the help of the Dulles approach, I climbed and descended like a roller coaster, searching for an altitude free of ice (which I found), and I had my first introduction to carburetor icing. But it wasn't until I stopped at Knoxville for fuel that conditions got to the point where I seriously had to consider terminating the flight. I had been flying on top at 10,000 feet. I was quite happy. But weather conditions being what they were, and my fuel consumption on the first leg being higher than I planned, I decided that I had to make another stop — even if it meant another descent and climb through the clouds. On my descent I was told that a Mitsubishi had reported moderate-to-severe icing, but I wasn't told when. Ceilings were about 1,500 feet. I reported light icing when I broke out and landed.

While the airplane was being topped off, I called flight service for the weather. They reported 1,000-foot ceilings and snow at Nashville. And I was told that at Knoxville there were reports of moderate-to-severe icing in the clouds. I began to worry again. Should I leave the solid earth to embark upon what might be a perilous flight to the tops of the clouds? I had to decide again whether I could enter the IFR system. At this point, I decided that with 1,500-foot ceilings I would have an easy enough time returning if the ice was as bad as reported. Once again I resolved to "have a look." I had options if even the worst happened, and I suspected that the Mitsubishi report was less than reliable. As I taxied for takeoff, I was informed yet another time of the report — Mitsubishi — of moderate-to-severe icing. I asked whether there were any other reports. "No," I was told.

I took off, entered the clouds, and devoted my scan to the instruments and the little thermometer probe. I arbitrarily decided that if the ice got about as thick as the probe I probably couldn't climb to 10,000 feet anyway. I would turn back. I did make it to 10,000. I reported light icing and was on my way.

Getting started on an IFR flight is still difficult for me, but my approach is different, now. If there is no reason why I should not take off — thunderstorms in the airport vicinity, icing with low ceilings, or any other situation that does not give the option to hightail it back to the airport — then I will probably opt to test the waters. Of course, this method does not dispense with the go or no-go decision, it just spreads it out over the course of the entire flight. I have always been able to make that sort of decision on the basis of tangible facts more ably than on speculation, but the responsibility not to shirk the decision becomes even greater in the air.

Although launching with the intention of evaluating conditions later has helped me to get into the IFR system more often, I ran into the problem of evaluating the weather, flying by instruments, and dealing with departure procedures all at the same time. Flying into a low ceiling just after departure is often unavoidable, but it compounds the risks of the first few moments of flight. Of the failures and mishaps that I have experienced (I'll include the time I left the door open on a Sierra), nearly all occurred in the first few moments of flight. I know that I am not unusual in that respect. That, coupled with my tendency to drop the instrument scan and a natural preoccupation with the weather, sets the stage for potentially serious consequences.

One tendency to which I am particularly prone is fixating

on one instrument on the panel or on conditions outside the airplane. Fixating is a common problem and a breakdown of one of the most basic elements of instrument technique, but even so I find that, especially after takeoff, it is very tempting to let my concentration lapse and just stare at an instrument or at the clouds outside. Few sights are quite as mesmerizing as the inside of a cloud. I'm not sure why that is, but often, with the intention of just checking whether I'm flying into darker cloud or light cloud, or to see how much ice is accumulating on the wing, I am easily drawn away from the business of flying the airplane. Part of the reason at the beginning of a flight, I suspect, is that I have not yet solidly established cockpit routine.

Another common problem is in-flight distraction. If something is going to go wrong, pilots often discover it in the first few moments of a flight. I find that when I have goofed up in preparation, I generally catch on to the resulting problems during climbout. While sitting on the ground preparing for takeoff, I may have set incorrect frequencies for navaids or for departure. So there I am, looking for a frequency and having to set it while climbing into instrument conditions. It is at such times that the old command, drummed into my head after hundreds of thousands of hours of lessons (or so it seems), comes back to me: "Fly the airplane, stupid." Every instructor with whom I have ever shared the inside of a cockpit has prodded me with those words, and I remember them when circumstances temporarily cause me to lose sight of my priorities in the airplane. It is especially true in the first phase of flying.

If my experience is any indication of that of other new IFR pilots, getting into the IFR system can be fraught with concerns about the weather, about the first moments on instruments, and about the airplane. I find it helpful to try to keep the extraneous concerns to a minimum. It is important to settle down into the IFR groove, to get a feel for how you and the airplane are doing, and then to move into the pattern of flying the airplane, managing navigation and communication, and planning the next phase of the flight. Of course problems and unforeseen situations may intervene, and if they do, there is no choice but to deal with them as systematically and efficiently as you know how. But even when they do, I think it is a mistake to neglect the need to adjust to the IFR environment. Sights and sensations are different when flying through clouds, and to ask an immediate transition of yourself is a tall order indeed.

I've found that the first phase of an IFR flight is one of the most important of an entire trip, but for some reason, it is one of the elements of a flight that my instructors never really discussed much except to say, "Don't get into conditions over your head." That warning, at best a statement of the obvious, never did me much good. What did help me, though, was talking to other pilots and giving their experiences (often similar to mine) some thought. What seems to me to be the best conclusion is this: If the weather is good enough to get started, then do it. Work from there, and use your judgment as to whether conditions stay good enough to continue. I worry sometimes that I might get into weather that is "over my head," but this philosophy has enabled me to cross the threshold between sitting on the ground and actually getting into the IFR system.

SECOND PART   Richard L. Collins

At the beginning of an IFR flight we set in motion a chain of events; once we lift off we are committed to completing the flight. It might not be the flight that we started out to complete — it could be shorter or longer — but all factors have to be dealt with before landing successfully. Two airline accidents and some general aviation accidents offer clear insight into some of the thought processes in a departure that is fraught with problems.

The first example comes from the Pan American 727 that crashed at New Orleans. There was a thunderstorm over the departure path. The aircraft lifted off and then struck trees 2,376 feet from the departure end of the runway and crashed.

To set the scene, consider the weather: New Orleans was under the influence of a high-pressure center located about 60 miles off the Louisiana coast. There were no fronts or lows in the vicinity on this typical midsummer day. There were no sigmets current for the New Orleans area, though one was current covering thunderstorms in an area beginning 60 miles southeast of New Orleans. The weather as reported a few minutes before the flight started its takeoff roll was: "measured 4,100 feet overcast, visibility 2 miles, heavy rainshowers, haze; wind 070 at 14 knots gusting to 20 knots. Altimeter setting 29.98 inches mercury. Remarks, cumulonimbus overhead."

An airline pilot who departed about ten minutes before the Pan Am flight stated that his radar showed a cell over the airport. Another airline pilot elected to depart on Runway 19 (Runway 10 was active and was used by the Pan Am 727) seven minutes before the accident. According to this captain, thunderstorms were all around the airport; one was east-northeast of the airport, and numerous cells were to the south, southwest, and west between 5 and 20 miles from the airport. The captain stated that the largest radar echo was east-northeast of the airport, and that the cell contoured when he switched to the contour mode on his radar. The captain testified that the gradient in this cell "was very steep." (A steep gradient is indicative of a strong storm.) About halfway through the takeoff on Runway 19, this aircraft began to drift to the right, and wind shear was encountered on initial climb. (They reported the shear to departure control, but the report was not passed along to the tower.) While only a few witnesses to the Pan Am accident reported thunder and lightning, the National Weather Service reported an intense cell with a maximum top of 49,000 feet in the area less than thirty minutes after the airplane was lost.

Go from that background to the flight deck of the 727. Only weather and takeoff-related conversation is included. Pertinent conversation between controllers and other aircraft is also included. Conversation between the Pan Am crewmembers is identified by crewmember only. The precision of text in brackets is questionable.

SOUTHWEST 860: Southwest 860, any chance of 28?

GROUND CONTROL: Ah, Southwest 860, that's ah, negative, unable at this time due to inbound traffic to 10.

USAIR 404: Ah, ground, USAir 404, we're still at the gate, ah, any chance of 1?

PAN AM FIRST OFFICER: With this thing, any more than 1 knot of tailwind, and we wouldn't be legal.

GROUND CONTROL: USAir 404, Runway 1 is noise sensitive for departures, advise your intentions.

USAIR 404: Roger, we'll have to look at the weather.

PAN AM: Clipper 759 taxi, and we need Runway 10.

GROUND CONTROL: Clipper 759, roger ah taxi to Runway 10, amend initial altitude 4,000, ah, departure frequency will be 120.6.

PAN AM: 20.6 and 4,000, Clipper 759 what is your wind now?

GROUND CONTROL: Wind 040 at eight.

PAN AM CAPTAIN: It'll be a heavy, ah, takeoff so if we have to take, ah, if we have to abort for any reason, you'll have the throttles, get all we can out of 'em now so if we bust one before *V* one, we'll stop and we'll ah stand on the brakes, Don. If it's past *V* one, go ahead and start dumping Leo.

CITATION ONE MIKE TANGO: Ah, Mike Tango, what's that wind doing now please?

GROUND CONTROL: Wind, ah, 060 degrees at one five, peak gusts 25, low-level wind-shear alert at, at northeast quadrant 330 degrees at 10, northwest quadrant 130 degrees at 3.

SOUTHWEST 860: Ground Southwest 860, with the present wind conditions we're requesting 28 for departure.

GROUND CONTROL: Southwest 860 roger, see what we can work for you.

CITATION ONE MIKE TANGO: And ah ground 31 Mike Tango is also requesting 28.

PAN AM CAPTAIN: Now we might have to turn around and come back.

PAN AM: What are your winds now?

GROUND CONTROL: Winds now 070 degrees at 17 and ah peak gust that was ah 23 and we have ah low-level wind-shear alerts all quadrants. Appears the frontal passing overhead right now. We're right in the middle of everything.

PAN AM CAPTAIN: Let your airspeed build up on takeoff. . . . The [wind's] going to be off to the left . . . not much. . . . I don't understand why these guys are requesting 28.

PAN AM FIRST OFFICER: I don't either. . . . [must be sittin' there look' at a windsock.]

PAN AM: Clipper 759 is ready.

TOWER: Clipper 759 maintain 2,000, fly runway heading, cleared for takeoff Runway 10.

PAN AM: Maintain 2,000, runway heading, cleared for takeoff Runway 10, Clipper 759.

TOWER: Eastern 956, Moisant tower, cleared to land Runway 10. And, ah, Eastern, the wind 070 at 17, heavy DC-8 er ah heavy Boeing just landed said a 10-knot wind shear at about 100 feet on the final.

EASTERN 956: Thanks very much.

PAN AM FIRST OFFICER: We're cleared for takeoff.

PAN AM SECOND OFFICER: Looking good.

PAN AM CAPTAIN: A slight turn to the left.

FIRST OFFICER: Takeoff thrust. . . . [need the] wipers.

CAPTAIN: [VR]. . . . Positive climb.

FIRST OFFICER: Gear up.

CAPTAIN: [V two. . . . come on back, you're sinking Don. . . . come on back.]

This was followed by the sound of the ground proximity system — woop woop pull up woop — then by the sound of impact.

There was undoubtedly a thunderstorm affecting the airplane's departure path despite the lack of a sigmet. Based on the radar report made about twenty-five minutes after the crash, it developed into a strong storm, as if it wasn't strong enough when the Pan Am 727 went down.

The 727 lifted off about 7,000 feet down the 9,228 foot-long runway, reached an altitude of about 100 to 150 feet above the ground, and then began to descend. A Boeing Company analysis of the flight data recorder suggested that the aircraft was operating with a 20-knot headwind at the time the captain called "positive climb." By the time it was 100 feet high, it was operating with a 20-knot tailwind as well as a downdraft velocity of almost 30 feet per second, or 1,800 feet per minute. Given that, the airplane would not have had the performance to fly level, especially after a maximum-weight, hot-day takeoff.

YOU ARE THERE

Put yourself in the cockpit of an airplane in a similar situation. If it's a light airplane, say a Bonanza, consider the runway to be a couple of thousand feet long — as minimum to a Bonanza as Runway 10 at New Orleans was to the 727. The choice is between taking off toward a storm, into the surface wind, or away from the storm, downwind, or waiting a while. Which would you do? This is an important part of getting into the IFR system.

In an airchair it's easy to say, "I'd wait a while." But only a pilot who is also a student of weather can tell you why it would be absolutely imperative to wait.

Given a surface wind from an easterly direction, a takeoff to the west would be impossible. In the 727, for example, a crewmember said it wouldn't be legal with but one knot of tailwind. The airplane rolled 7,000 feet taking off *into* the surface wind. Certainly you'd hardly want to fool with a downwind takeoff in a loaded Bonanza on a 2,000-foot strip. So, scratch one alternative.

If the decision has to be fly, the takeoff to the east appears the only alternative. The fact that only a few of the witnesses reported thunder and lightning might have suggested that the rain was just from a heavy shower instead of a thunderstorm. This can be a dangerous illusion. In the southern U.S. in July, isolated heavy rain comes almost exclusively from cumulonimbus clouds. The fact that the National Weather Service wasn't calling for thunderstorms had absolutely nothing to do with it. What you see is what you get. At New Orleans, one of the preceding pilots described what could only be considered a heavy thunderstorm over the departure path.

The structure of a thunderstorm is relatively simple — air flows in around the edges to form the updraft, and there is a downdraft in the center. That's elementary meteorology. It's also rather common knowledge among pilots who fly in the semitropical southern U.S. summers that stationary or slow-moving storms can be quite mean.

In taking off toward a storm, with the surface wind coming from the direction of the storm, you know that you are in the outflow of the storm. Basic. Headed toward the storm and climbing you'll ascend at some point into the inflow. Also basic. That would mean flying from a headwind to a tailwind in the climb. That can result in a serious decay of airspeed — the heavier the airplane the worse the decay because the mass of the airplane makes it react to the new wind more slowly. There would also be the downdraft to contend with, likely strongest in the area of heaviest rain. Finally, it has been hypothesized that the ability of a wing to generate lift is degraded by rain flowing over the surface. Any way you look at it, taking off toward an active cell stacks the deck against a successful departure. In the case of the 727, the low-level wind-shear alert system (which is nothing more than anemometers at various locations) gave a strong indication of the shallowness of the northeasterly surface wind. The breeze was actually from the northwest on one corner of the airport. And remember, on initial climb there's nothing to use against a sinking spell; the power is likely already at the limit.

In such a situation, a moment spent visualizing the air into which you are about to fly can do wonders for health and happiness. The basics of a thunderstorm outlined what could happen, and the crews of at least two other airplanes gave the subject thought and requested a takeoff in the opposite direction. The crew of the 727 did consider why the other crews were asking for a different runway, but because of the weight of their aircraft and the surface wind, they had no choice other than to go on 10 or not go at all. Being very experienced pilots, they had undoubtedly flown through what they considered similar circumstances before. But one thing that you learn by flying a lot of hours over an extended period of time is that there are no hard and fast lessons found in experience — only valuable generalities. The fact that one situation appears similar to another doesn't count. Each must be dealt with by weighing all the facts and basing action more on a strong knowledge of factors affecting the flight than on experience with a cloud of similar appearance.

SNOWSTORM

Consider next an event in which the pilots weren't experienced with the conditions that existed as they tried to get into the IFR system. The crew of the Air Florida 737 that crashed right after takeoff at Washington National in a snowstorm hadn't flown a lot in weather such as existed at National that day. The following transcript gives some insight into the conduct of the flight up to the time it hit the 14th Street Bridge and crashed into the Potomac River. Palm 90 is the radio call of the aircraft in question; only pertinent portions of the transcript are included. Where the transmission is identified only by the crewmember, it is inter-cockpit conversation. Items in brackets may not be precise.

FIRST OFFICER: Boy, this is [expletive], it's probably the [expletive] snow I've ever seen.

CAPTAIN: . . . Go over to the hangar and get deiced.

FIRST OFFICER: It's been a while since we've been deiced. . . . That Citation over there, that guy's about ankle deep in it.

CAPTAIN: Tell you what, my windshield will be deiced, don't know about my wing.

FIRST OFFICER: Well all we really need is the inside of the wings anyway, the wingtips are gonna speed up by 80 anyway, they'll, they'll shuck all that other stuff [sound of laughter].

CAPTAIN: [Gonna] get your wing now.

FIRST OFFICER: D' they get yours? Can you see your wingtip over 'er?

CAPTAIN: I got a little on mine.

FIRST OFFICER: A little. This one's got about a quarter to a half an inch on it all the way. . . . Look how the ice is just hanging on his, ah, back, back there. See that? See all those icicles on the back there and everything?

CONTROL TOWER: Now for Palm 90, if you're with me you'll be going out after, ah, the red DC-9 Apple type.

AIR FLORIDA: Palm 90.

FIRST OFFICER: Boy, this is a, this is a losing battle here trying to deice those things. It [gives] you a false feeling of security, that's all it does.

CAPTAIN: That, ah, satisfies the Feds.

FIRST OFFICER: Yeah. . . . Slushy runway, do you want me to do anything special for this or just go for it?

CAPTAIN: Unless you got anything special you'd like to do.

FIRST OFFICER: Unless just take off the nose wheel early, like a soft field takeoff or something. I'll take the nose wheel off and then we'll let it fly off. Be out the 326, climbing to five; I'll pull it back to about one point five five, supposed to be about one six, depending on how scared we are. [Sound of laughter.]

CONTROL TOWER: Palm 90, taxi into position and hold. Be ready for an immediate.

AIR FLORIDA: Palm 90, position and hold.

CONTROL TOWER: Palm 90, cleared for takeoff.

AIR FLORIDA: Palm 90, cleared for takeoff.

CONTROL TOWER: No delay on departure if you will, traffic's two and a half out for the runway.

AIR FLORIDA: Okay.

CAPTAIN: Your throttles.

FIRST OFFICER: Okay.

CAPTAIN: Holler if you need the wipers.

CAPTAIN: It's spooled.

CAPTAIN or FIRST OFFICER: Ho . . . whoo . . . Really cold here.

FIRST OFFICER: Got 'em?

CAPTAIN: Real cold. Real cold.

FIRST OFFICER: God, look at that thing. That don't seem right, does it? . . . Ah, that's not right.

CAPTAIN: Yes it is, there's 80.

FIRST OFFICER: Naw, I don't think that's right. . . . Ah, maybe it is.

CAPTAIN: Hundred and twenty.

FIRST OFFICER: I don't know. . . .

CAPTAIN: *V* one . . . easy . . . V two.

CAPTAIN: Forward, forward. . . . We only want 500. . . . Come on, forward . . . forward . . . just barely climb.

CAPTAIN or FIRST OFFICER: [Stalling] we're [falling].

FIRST OFFICER: Larry, we're going down Larry.

CAPTAIN: I know it.

The crash followed that last transmission.

This was a much discussed accident. There apparently

were erroneous engine indications, caused by icing. The engine instrument indicators probably prompted the references to cold and the first officer's questioning of whether or not things were right. These gauge indications prompted the crew to set the power much lower than it should have been for takeoff. The captain's failure to reject the takeoff when his attention was called to the anomalous instrument readings was listed by the NTSB as one of the probable causes. The decision to take off with ice and snow on the airframe was another one of the probable causes, along with a known inherent pitchup characteristic of the 737 when the leading edge is contaminated with even small amounts of snow or ice.

MORE ICE

A Beechcraft Super King Air was involved in an ice-related accident that also offers a pertinent lesson on the beginning of an IFR flight.

The aircraft flew from Texas to Arapahoe County Airport near Denver, arriving just after 10:00 A.M. On arrival, the crew gave a fuel order with special instructions not to exceed the specified amount of fuel. Soon after landing, the captain of the aircraft called the FSS for a weather briefing, to cover a return flight scheduled for about 1:30 P.M. The briefer told the pilot of light rime icing reports and a forecast for moderate mixed icing in the area by afternoon. Having just arrived, the pilot was familiar with the conditions that existed.

The crew returned from lunch about 1:30. Snow had been falling, and they asked a ramp employee to start deicing the aircraft. The snow was brushed from the wings and the tail. The employee said the snow was wet and that it came off the aircraft easily, leaving a beaded film of water. It was estimated that from one to two inches of snow remained on the fuselage of the aircraft. The surface temperature was about 32°F.

There was no record of the pilot getting an updated weather briefing, which would have included a forecast for occasional severe icing. The passengers arrived at 2:10, and the ramp employee stated that he overheard the captain remark to the copilot that they were going to "have to burn off ten to fifteen gallons of fuel during the taxi." In fact, the National Transportation Safety Board concluded that the airplane was about 600 pounds over its takeoff gross weight. In addition to the crew, there were eight passengers aboard.

The crew received a prompt clearance from air traffic control and took off at 2:34. One witness stated that the aircraft lifted off about 4,700 feet from the end of the runway; other witnesses stated that its initial climb angle was "shallow" when compared with other aircraft. However, the NTSB later computed that the airplane's initial climb rate was normal for the weight.

The climb apparently didn't stay normal for very long. About ten minutes after takeoff, the copilot told departure control, "Okay, we would like to go back to Arapahoe County. We're getting [a] little too much ice up here." The aircraft was equipped and approved for flight in icing conditions; there was no mention that the equipment wasn't operating properly.

At the time the copilot requested a return to Arapahoe, the aircraft was 14 miles from the airport at an altitude of 12,700 feet. (Field elevation is 5,872 feet.) According to NTSB's computed data the aircraft had climbed at a decreasing rate to 12,800 feet where it began descending even though it was cleared to a higher altitude.

About a minute after the copilot said they wanted to go back to Arapahoe, he radioed, "We want to go to Stapleton now." Stapleton, the large airline airport serving Denver, was about 25 miles northwest of the flight at that time.

The controller commenced vectors and then cleared the airplane to a VOR, to maintain 11,000 feet. Next came a vector for sequencing. Then the crew asked for the Colorado Springs weather, which was sky partially obscured, 500 overcast, visibility 2½ with light snow showers and fog. The next transmission from the flight was, "Okay, we were asking below 11,000, we can't hold it here at eleven."

The controller asked if they still intended to land at Stapleton; the answer was affirmative. The controller then cleared the flight to 10,000 and told the crew that was the lowest possible at the time because of traffic. Then, just over five minutes after first requesting a return to Arapahoe and but about fifteen minutes after takeoff, the crew radioed, "Okay, we'll declare emergency if necessary, just get us straight to the runway." The controller cleared the aircraft to 8,000 and issued a vector that would take it to the final approach course. The crew responded, "Gonna be descending all the way." The controller asked for a repeat of that transmission and the crew responded, ". . . gotta come right on down, you better get us to the nearest airport." The controller informed them that they were 21 from Stapleton and asked if they would be able to make it. The reply was: "Naw, get us to the nearest airport. We gotta come on down." This was the last recorded transmission from the flight, which crashed at about 2:52, or about eighteen minutes after takeoff.

This accident clearly illustrates that deicing equipment isn't the answer to every question, and that a decision to "take a look" must be quickly reversed if things go sour. The icing was indeed heavy in the Denver area that day, and while other aircraft were operating, none was trying to climb at the same time the King Air accumulated its debilitating load of ice. An

MU-2 on approach to Arapahoe County at about the time of the accident missed twice because of ice on the windshield; the pilot declared an emergency and landed at Stapleton. The crew of the MU-2 reported an accumulation rate of $1/8$ to $1/4$ inch of ice every forty-five to sixty seconds.

Another King Air was operating in the area at about the time of the accident. The pilot of this aircraft reported accumulating about $1^1/2$ inches of ice. This aircraft descended and made a normal landing at Arapahoe County about eight minutes after the other King Air crashed.

The NTSB concluded (among other things) that the rate of ice accumulation on the aircraft lost was estimated at 0.14 to 0.19 inches of ice per minute. They also concluded that the wings lost lift because of the rapid rate of ice accumulation on unprotected surfaces — specifically on the underwing surfaces aft of the deicer boots. In its probable cause statement, the Board cited the flightcrew for operating an "over-gross-weight aircraft at high angles of attack in severe icing conditions."

Things happened rapidly to this flightcrew. The lesson for the rest of us is in the fact that deicing equipment, if we have it, is something to use when fleeing ice. If there is a very rapid buildup, it might well overpower the deicing equipment — especially in a climb when ice can accumulate behind the boots. It has always been my feeling that airplanes shouldn't be "approved" for flight in icing conditions because this gives tacit encouragement to a pilot who is tempted with pressing on in severe conditions. Certainly the equipment should be tested and pilots should be advised that it has been tested and will effectively remove ice from protected surfaces. But to parlay this into what is a blanket "approval" to fly in icing conditions is a bit much.

The NTSB also listed the flightcrew's failure to obtain a current weather briefing as a probable cause. Certainly we should always make that last minute assessment of the weather. Some pilots might feel that an earlier flight and a look at what appears to be an unchanged sky would suffice. Ice and thunderstorms, though, are two things that we can't try too hard to learn about. Whether or not the crew would have attempted the flight had they checked the weather is an unknown.

A HASTY DEPARTURE

The importance of proper preparation and a methodical approach to getting into the system is emphasized in the NTSB accident report on a Citation II that crashed while departing from a fog-shrouded airport.

The Citation was approved for single-pilot operation. On this morning, the president of the company that owned the airplane was flying, and he was apparently in a hurry. He called from his home to the flight service station for a clearance at 9:09; he got one with a void time of 9:30. The pilot then drove to the airport, arriving between 9:20 and 9:25. He loaded the baggage and boarded the airplane, along with two passengers. Both engines were started and the airplane taxied to the departure end of Runway 28. The pilot stopped the airplane in position on the runway. It remained there for a half a minute or so and then started to take off — only about two minutes after the engines had been started.

The takeoff appeared normal to qualified witnesses. The airplane disappeared from sight when it was 20 to 25 feet above the runway. A person near the airport heard a jet fly over his house and shortly afterward heard an explosion. The airplane had crashed in woods about 1.75 miles north of the airport. At the time of impact the airplane was on a heading of 120 degrees and was in a 30-degree nose-down attitude and 90-degree bank to the left.

Visualize for a moment the situation. The pilot took off to the west and crashed north of the airport on a southeasterly heading. The heading for the destination would have been 030, and the National Transportation Safety Board came up with an interesting scenario of how the airplane arrived at the scene of the crash. The Board theorized that the pilot started a right turn after takeoff but never rolled out on a heading of 030. "Rather, for some reason, the airplane might have continued in a progressively steeper right bank and tighter turn to the right which developed into a continuous roll to the right. The airplane could have rolled through 90 degrees to the right to the inverted position and continued to a 270-degree right bank attitude, or, a left vertical bank of 90 degrees, when the airplane hit the ground." Whatever, it was clear that the airplane was out of control.

Why?

The NTSB noted that the flight instrument system of this type airplane requires a maximum of three minutes, depending on temperature, to become operational. The electrical power had been on for but two minutes when the takeoff roll started. Additionally, on this particular airplane, a difference in times required for the horizontal situation indicators to come up to speed was reported by another pilot who had flown this airplane. The person flying at the time of the accident had used the copilot's HSI for heading information the previous day because the pilot's HSI was slow in becoming operational.

According to the report, the pilot's attitude indicator could have been working properly. "The takeoff was probably made with the pilot flying the airplane manually, using attitude information provided by the pilot's attitude director indicator, but most likely using the copilot's horizontal situation indicator for heading information." The probable cause was a loss of control "as a result of the pilot's use of attitude and heading instruments which had not become operationally usable and/ or his partial reliance on the copilot's flight instruments which resulted in an abnormal instrument scan pattern leading to the pilot's disorientation. Contributing to the accident were the pilot's hurried and inadequate preflight procedures."

This pilot was described by acquaintances as a strong-willed, aggressive individual who had total confidence in himself as a pilot and businessman. Pilots who had flown with him said he was a very skilled pilot although he sometimes violated certain aviation safety practices. Four persons said they had been in the airplane with him when he had rolled the airplane.

One thing we have to learn early-on and continually remember is that instrument flying is totally demanding. To make any exception in haste carries with it a considerable risk. And flying without the flight instruments in a proper operational condition increases the risk dramatically.

The basic instrument systems in most singles and smaller twins have a disorienting characteristic that is worth considering here. The stage is set for this if you run the engine for a short while and then shut it down, as you might do when positioning the airplane to load passengers or baggage. If you then start it again within a few minutes, before the gyro in the artificial horizon spins down completely, some instruments tend to spin back up but not give a proper indication immediately. They remain where they were when the airplane was restarted. They might show, for example, the airplane in a 20-degree bank. If a pilot were to hurry into an instrument takeoff, he might find this disorienting indeed. It is not something that could or should catch an alert pilot — the indication will go slowly from incorrect to correct and will be obvious when it is incorrect — but could be a factor if a pilot did not carefully check everything before launch.

MIND-SET

Here we need to contemplate what might be considered a mind-set. Once the decision is made to go, everything proceeds toward that end. We see other people coming and going, which reinforces what we consider to be a good decision. After waiting a while for a clearance, we might not be prone to throw it in the can and taxi back to the ramp because of some doubt about equipment or conditions. Whether the problem is ice on the airframe, a recalcitrant magneto, flaps that don't work, forgotten charts, or a bellyache, it's human nature to think: "Can't I deal with this? Can't I keep going and make the necessary compensation for a less than ideal situation?"

If confession is good for the soul, we can all cloak ourselves in a lot of goodness in this area. I certainly recall situations where I have tried to rationalize away bad things — and a lesson is usually learned as a result.

One was related to ice. It happened earlier, and I thought a lot about it after the Air Florida accident report was released.

I was on the ramp at Indianapolis for only long enough to top the tanks. Heavy wet snow was falling, and I just knew that I'd get back out before enough snow accumulated on my airplane to matter. After paying the fuel bill and going back to the airplane, though, I knew I was wrong. There was about a half-inch of snow on the airplane. Nothing to do but get it off before flying.

The FBO had a hot glycol truck, but it was away hosing

down a bigger fish. "Be a while before it gets back." Because I'm often where a pilot is left to his own deicing devices, I carry my own equipment and decided to use it this day. My gear consists of a windshield broom and at least seven cans of windshield deicer. (I don't recommend the latter because some people say that it is harmful to paint and windows. It does work on snow and ice, though, and by careful use I've been able to avoid any obvious harm to anything.) I swept away as much snow as I could and then squirted the wings and tail with deicer to achieve that clean and slick look. It took only about ten minutes, and I felt self-righteous about the conscientious job of deicing. "This lad," I thought, "will not take off with snow or ice adhering to the flying surfaces."

I hopped aboard, reeking of deicer, got my clearance, and taxied out. The heavy snow continued to fall.

There were two airplanes ahead of me, an airline 727 and a business jet. The bizjet had been there for a while, and I noticed some snow on his wings and fuselage, but he took off without incident.

The airline jet was next to go after a bit of delay. It disappeared into the snow before rotating. Then it was my turn: "Taxi into position and hold. . . ."

I told them I'd like three minutes behind the departing 727 because of wake turbulence . . . if I didn't get off as well as usual. That was okay — no inbounds — and by the time I was cleared to go, probably twenty minutes had elapsed since I completed my deicing job. I knew there had to be some snow on the wings, but the temperature was only a degree or so below freezing, and from experience I knew that the deicer would help impede formation of new deposits — at least for a little while. It being a high wing, I couldn't see the snow, but there was some visible on the tail. I felt sure there would be no problem, but I resolved to abort the takeoff if the airplane didn't lift cleanly into the air and commence climbing at its normal 70-knot rotation speed. That plan insulated me from any possible problem, I thought.

The takeoff roll was normal so I gave no thought to aborting. At 70 knots, I gave the requisite tug on the wheel and the nose of the airplane obediently lifted. Then the airplane flew cleanly away, all according to the script. As I lifted off, I heard conversation about the airliner that had preceded me needing to return and land. He was apparently having some sort of a problem with his flaps — like they probably wouldn't retract because of ice and snow.

As I started to turn on course, my airplane suddenly became wing heavy. It took about 20 degrees of wheel travel to keep the wings level once on heading. I thought some rather dumb thoughts. Had the FBO filled one tank and left the other empty? No, the gauges were full, and I had looked into the tanks. There was a fleeting thought about wake turbulence, but that was illogical. Finally I admitted to myself that the flying surfaces had probably had some snow on them, and that one wing had shucked its load while the white stuff was clinging tenaciously to the other one. There couldn't have been more than a quarter of an inch on there, but the difference in the work done by the two wings was significant — and impressive to me.

The lesson, on any IFR departure, was to evaluate the conditions immediately prior to departure and base the decision to go on what's happening then. A lot can change between the time an airplane is deiced or the weather is checked and twenty or thirty minutes or even an hour later, when you are finally ready to take off. If things don't look right at the last minute, they probably are not right. It's critical to abort any IFR takeoff in which everything doesn't act, sound, and look right. It's a time when you are taking a lot of machinery from dormant to a highly active state, and you sure want to investigate any abnormality. A hunch that things are okay, based on either a guess or some previous experience, just isn't enough.

ICY DAY

There's no way to lock experience out, though, and a flight from Wichita to Phoenix in a P210 illustrated how the weather information system can combine with a previous experience to make one nervous about a routine departure.

I had a colleague with me on this flight. He had flown into Wichita the evening before in his Mooney and had garnered quite a load of ice on the way in. I felt guilty about that because I told him on the phone that I hadn't accumulated any ice arriving in my 210, despite the fact that all the forecasts were calling for the bad stuff. He had originally planned to take his Mooney on to Phoenix but, based on information available from the FSS the next morning, he opted to ride along in my airplane, which was equipped with deicing equipment.

Even in the deiced airplane I put a lot of thought into takeoff alternates. The most current radar report showed 60 percent coverage of precip for the first 100 miles. Moderate icing was forecast. The surface temperature was 34, so the ice would start quickly, and if it was heavy ice the climb would be affected almost from the beginning. Tops of the precip were shown as 25,000 feet, and there was an old pilot report of freezing rain from the surface up to 6,000 feet. (We later figured out that this was probably the report given by my Mooney pilot-passenger the night before.) It was snowing out as far west as Amarillo, where the forecast called for occasional 200 obscured in heavy snow. The takeoff alternate had to be Wichita, the departure point. The current weather was 700 overcast and

6, with that value forecast to continue through the morning. My plan was to take off and have a look; if things weren't good it would be a return to Wichita and land. The wind aloft was light at 12,000 feet but strong and southwesterly at 24,000 so we had good options there. Go west low and if it became necessary to return to Wichita we could climb up into a strong tailwind — if the airplane was relatively ice free.

A lot of thought went into the Mooney pilot's decision not to go and into my decision to go. We each felt as if we had made a good choice. I was looking forward to an interesting flight.

As is often the case, we were on top at 10,000 feet. There was no ice. The air was smooth. The only change in plan came at Tucumcari where we had intended to stop for fuel. The FSS person there suggested that we go on to Albuquerque if we had the fuel to do so. The Tucumcari airport was covered with ice and snow; we hadn't been advised of this during the weather briefing.

My Mooney pilot was, naturally, chagrined that he had left his airplane behind. It would have been an easy flight for him. But it's an illustration of how experience isn't always a positive factor in dealing with weather. The experience of that flight could lead the unwary to launch a mission impossible because of thoughts of a day when a trip was canceled due to incorrect inclement forecasts. Again, you have to take them one at a time, for what they are. And that day, we had an alternate plan to use if the weather had been worse instead of better than forecast.

A TIME FOR GREMLINS

Even if an airplane is properly preflighted, mechanical or electronic things can go wrong. And even though there are no statistics to bear this out, I think the chances of something failing are higher during a departure, in the first minutes of flight, than later on. If this is where a new instrument pilot like Bradley has most of his problems, it is also where airplanes have theirs. The takeoff is where you push things to the limit. Additionally, there's a gremlin who lets things work until they are really warmed up (hot in the case of some avionics) and then makes them go fizzle. An event that occurred one night underlined the desirability of thinking about all departure possibilities.

The single-engine retractable was on a cargo flight. Right after punching into a low overcast, there was a flash and all the electricity went away. No radio, no lights. Nothing was left. When an alternator quits, you have the residual energy in the battery to parlay into a landing. These lads had nothing but a flashlight with relatively old batteries.

They dead-reckoned to where they thought the weather was better and made a letdown. Unable to find an airport, they finally decided to land the airplane in a field. The landing was a relatively successful event though they had no flaps and no lights. (The landing gear was extended with the emergency system.)

I learned several things from this event. One was the overwhelming value of the independent battery-operated transceiver that I keep talking about. Had they had one of those, it would have been a simple matter to get a radar approach back into the airport they had just departed from. Another lesson was in the failure. You *can* lose the whole electrical system. It has happened in singles and twins, business jets and airliners.

There might have been a clue to an impending problem. There have been instances of the starter not disengaging on piston engines. When this happens, the starter is turning at a very high rpm rate — much higher than it is designed for. It'll rather quickly burn out and cause a short to move through the electrical system backward — through the battery. There is nothing to protect the system from this, and the short aces everything almost instantaneously. The length of time it takes to burn out a starter roughly corresponds to taxi, runup, takeoff, and climb to a few hundred feet. Some airplanes have a warning light to indicate that the starter hasn't disengaged; on others, a continuing high charge rate on the ammeter can be a sign of a hung starter.

RATIONALIZATIONS

When something seems wrong with an airplane, there's no way a pilot can stay completely away from a little rationalization — the key to success is in not kidding yourself for an extended period of time.

My son Richard was flying my P210, and I was in the right seat leaving central Florida for the Northeast. The climb was punctuated with level-offs, vectors, and requests for a better rate of climb. As we were climbing through 11,000 feet, Richard make a remark about the lousy climb rate and about not being able to keep the EGT as low as I like it in a climb.

My first thought was gasoline. At times, I've noticed a

tendency toward vapor problems when leaving Florida. I always charged it to the temperature of the fuel and have turned on the auxiliary fuel pump to smooth the flow of fuel to the engine. This day I did just that. But the rate of climb remained low. Instead of going to Flight Level 190, as we had filed, I told Richard to tell the controller that we would like to level at 15,000 feet. There I started fiddling with the mixture and found that the EGT would have peaked at a much higher temperature than normal. The engine wasn't running as smoothly as usual, but then it wasn't running rough.

I had had a magneto failure in the airplane a few months before. That time, there was a noticeable irregularity in engine operation. On this flight I sat trying to convince myself that the abnormal EGT reading was caused by fuel quality and the poor climb, and that the below-par cruising speed was a result of subsidence — the air flowing into a rapidly developing low center off to our right, causing a large scale settling of air in the area.

The weather beneath was lousy. Jacksonville was at IFR minimums. But as we moved on, the devil on my shoulder finally awakened. Something is not normal, the weather ahead at Savannah is a bit above minimums, there is a good shop there, and there's an extra mag in the baggage compartment. The thought of the extra mag convinced me that I really knew what was wrong, I just didn't want to admit it. Decision time: Stop at Savannah. (If you wonder why I didn't check the mag aloft, it's because a dead mag would result in a depressurization of the cabin, and then a repressurization when the engine restarted — to say nothing of probable harm to the exhaust system.)

Anyway, we got clearance to land at Savannah, shot an approach just about to minimums, found that the right mag was dead, had the spare installed by Butler Aviation, and were on our way in an hour. In retrospect, the stop should have been made in Jacksonville.

BACK TO NORMAL

Fortunately, most departures are normal, without pilot or airplane "events." But there can be strong challenges even during a normal departure.

The purpose of the departure is to get the airplane into flyable air and into the air traffic control system. The challenge is to do this as smoothly as possible.

When leaving an airport with radar service, the departure can be a matter of being led by the hand through the procedure. "Turn to a heading of 030, maintain 3,000 feet."

That clearly defines what to do and lets us program our autopilot (or ourselves) for the initial chore.

But how often have you forgotten the initial heading or altitude and had to refer back to the written clearance or ask the controller to say it again?

A pilot who hasn't forgotten much hasn't flown much. On more sophisticated airplanes it's easier: The heading goes on the heading bug and the altitude goes into the altitude preselect. But even in an airplane without the most sophisticated features there is often a way, or at least a partial way. An ADF with a movable card can serve as the altitude reference. Just set the altitude to which you are cleared at the top. When I traded my old ADF for an RMI slaved to the HSI, I lost that feature. It took me a couple of trips, and forgetting a few times, to come up with a new altitude reminder system. It turned out to be the true airspeed indicator. I set the first index mark on the assigned altitude (using the temperature scale on the indicator as my altitude reference). The number two nav can be used, and no doubt there are other things in the airplane. You can just write it down, or there is a simple nonelectronic heading and altitude reminder available — if you have an extra space in the panel for it.

Regardless, use what's available for an altitude and heading reminder. Altitude is probably the more important of the two because more time usually elapses between the time you are assigned a new altitude and when you reach it. For some reason, I have less trouble remembering headings, and even if you miss by 10 or 20 degrees you are on a track that could be a reasonable product of the heading, given the usually unknown wind aloft. Altitude is a more precise business. If you err, the mistake is in thousands, and if you are at the wrong thousand it's a serious matter. With altitude encoding, the controller would catch the error, but it might take a moment. Things can get close before the error is caught.

If the departure is from an airport without radar, you are on your own. As we discussed in the previous chapter, a knowledge of terrain is important. This is covered in the departure procedures section of the instrument approach chart. If there is a terrain or obstruction reason for departures to be restricted, it will be there. If no procedure is specified and you have some doubt about terrain, flying the instrument approach backward is a way to stay on a published route at a published altitude. Here, though, you are on your own until you get up to the minimum descent altitude, which can be rather high in some mountainous-area approaches.

At times we leave airports where there is no published approach. Here it is strictly up to the pilot to develop a procedure that eliminates terrain or obstruction risks. A current sectional chart is a necessity. Even if the aircraft is being operated clear of clouds to a published altitude, it's worth establishing a departure procedure. This is especially true at night.

SILENCE

Even though there are procedures to use in case of communications failures, it's common to become uneasy when the controller does not answer. This can be particularly disconcerting on departure, when the clearance usually contains restrictions with no specific instruction on when and where the restrictions will be lifted.

I was taking off to the north at Little Rock, Arkansas, one day in a twin turboprop and was told to "turn left to a heading of 270 degrees, maintain 2,000, contact departure . . . control. . . ." I was not familiar with the avionics system in the airplane; the pilot in the right seat was more familiar but there were some features still new to him. I leveled at 2,000, on a heading of 270, and called departure control. No answer. I called again. Still no answer. Hmmmmm. I lived in Little Rock at the time and was familiar with the area. There was an obstruction sticking up above our flight level but a few miles to the west. Clearly we couldn't fly a heading of 270 at an altitude of 2,000 feet while figuring out what was wrong with the radios. The proper thing would have been to climb to a minimum safe altitude while sorting things out. As it was, we figured out which buttons to press and soon had things back to normal.

Initial confusion has gotten the best of instruments pilots, leading to serious accidents. A confused clearance, a communications problem, a navigational error — any number of things can divert a pilot's attention from operating the airplane. The result can be a loss of control. On departure, there's not much room for error or much time to recover from an unusual attitude. It is a demanding time and a bad place to allow confusion to hold sway. The key is in knowing the heading and altitude that gets the flight off to a safe start, and getting the airplane established on that first part of the clearance. If there is doubt about anything, it is best resolved *before* takeoff.

CHAPTER 4

Operating in the System

FIRST PART   Patrick E. Bradley

I've heard stories about pilots, good pilots, who, once they are established in cruise flight, flip on the autopilot, pull out a magazine or a crossword puzzle, and amuse themselves with news or games until beginning the approach. Like waiting in a dentist's office. I don't doubt that they adequately monitor the instruments, their position, and other vital elements of the flight. And I don't doubt that, in good weather, which is when these pilots find cruise flight boring, they are "safe" as long as they don't collide with another airplane. I, however, have never been able to "kick back," even during cruise. Mostly, it's because I've just never had the time.

After listening to proponents of several schools of thought on cruise procedure, I've finally cast my lot for one in which the pilot busies himself with the airplane, following set procedures whether in good weather or bad. One reason for this is that I don't think I've got the level of skill or experience, yet, to allow me to switch modes — informal for good weather and formal for bad — from one flight to another. Without establishing and following set procedures to handle the cockpit work load in all conditions, I think I'd be at a distinct disadvantage in poor conditions. No, I don't call flight service for weather updates when conditions are clear VFR and forecast to stay that way. And I don't ask ATC if it's painting weather when I can see 50 miles ahead, but I do spend a lot of time calculating fuel and distance problems, working out "what if" scenarios, and generally trying to create work that may come closer to simulating the cockpit work load I will face in actual IFR flying conditions.

More than anything, my procedure is an exercise that often serves no purpose other than to hone my skills for times when I'll have to handle a high cockpit load, even if I've got to simulate it. Cruise flight in good weather can be boring, but in actual IFR conditions, it is a critical time. Preparation and accuracy can make the difference in a successful and unsuccessful flight. An added benefit of the system that I follow is that it stays boredom, a demon that torments any pilot on an uneventful flight. Nothing slows a trip like inactivity, and nothing, besides jet engines, can speed it like activity. So my exercises are really far from useless, and, in the event of a heavy cockpit load, I can fall back on set procedures. I will know what to do next, because I've been through it all, or nearly all, before.

It's never possible to completely prepare for a flight through IFR conditions, though, especially when faced with demanding things like thunderstorms, or icing. Both phenomena can test the furthest limits of a pilot's skill. After watching highly experienced pilots, I know that to deal with the worst conditions effectively, you've got to muster all available resources, including the folks on the ground. Dealing with ATC, like dealing with any people, is an art. The more a pilot can perfect that art, realizing the limitations of the assistance that ground personnel can offer, the more effectively he will be able to deal with weather.

One of the most common requests that pilots have is for reports on thunderstorm activity. At times, such a request will be met with pretty explicit guidance on the best route to follow, and at other times, a controller will remain completely tight-lipped. It's understandable, because controllers may not want to, and shouldn't be expected to, take responsibility for a pilot's decisions regarding thunderstorm avoidance. At other times, the controller's radar scope might not be equipped to read precipitation reflections associated with thunderstorms. Once, flying through heavy showers, I asked what sort of weather the radar was painting along my route. Center responded that there was some thunderstorm activity in the area, but beyond that, they just couldn't tell; their radar was equipped to follow airplanes, not rain. I decided at that point that my best bet would be to descend, and I was cleared to a lower altitude. When I reached the altitude, I was switched to an approach controller, and when I called, was immediately given vectors, without even asking, around thunderstorm activity. I'm still not sure why one controller had a better picture than another, but the situation repeats itself often. One thing I have learned, though, is that not even the most decisive controller can see and feel what the pilot sees and feels. Controllers can help, but they should never be relied upon for the final answer. Their reports and radar information never come with a money-back guarantee, so pilots, as always, must depend in the end on their own assessment of the situation.

In the thick of nasty weather, a pilot often must leave the airway, or his assigned altitude, in search of better conditions, and controllers will almost always cooperate where they can. A pilot once told me a story of a time when he requested permission to divert around thundercells that he could see pretty clearly. Permission was denied, for one reason or another. In response, the pilot told ATC that he would have to cancel IFR then, because he couldn't fly the airway. He and ATC finally reached an agreement on a mutually acceptable diversion. I'm not sure that I would freely give up an IFR flight plan to fly around thunderstorm cells, but this is an example that, with a little give and take, pilots and ATC can generally accommodate each other. But in the end, it's the pilot's call. Decisions regarding the safety of the flight are in the pilot's hands, and in the end it is he, not the controller, who will be held accountable.

Even with all of the resources available during flights through storms and ice, it is difficult not to feel a certain amount of stress. One way I've been able to counteract stress is never to get myself into a situation from which I may not be able to return. Keeping options open and setting definite guidelines as to when I will use them has helped me to approach flying somewhat more methodically, reducing the unknowns that contribute to stress. Still, nerves are a part of flying through adverse conditions, and there are times when, instead of working for the pilot by heightening his awareness and reactions, they can fog his reasoning.

Flying along the northeast coastline over southern Connecticut at about 6,000 feet, I had successfully eluded some light icing conditions at 8,000 feet. Skirting the bottom of a cloud deck, thinking myself in the clear, and catching reassuring glimpses of the ground below, I punched in and out of the bottom of the clouds, droning along merrily. Although I was still on the watch for airframe ice, toying with the idea of descending lower, I noticed that the airplane was drifting out of trim and that the rpm were decreasing slightly. I increased power, making a mental note to keep an eye on the situation, and continued.

In a few moments I was back in the clouds, and the intensity of the rain was picking up. The ride was fairly smooth, though, and the engine seemed to be operating normally. By the time I looked back at the tachometer though, I could already tell by the airplane trim that I was still losing rpm. By that time I still hadn't figured out what the problem was, so I started to make contingency plans. Just as I was about to call ATC and tell them that I would be landing at the nearest airport, the nature of the problem dawned on me: carburetor ice. I put on full carb heat, and after a few sputters that brought my heart up to my throat, the engine regained normal power. Looking back, I think that stress must have played some part in my carburetor ice scenario. If someone had just walked up to me on the street, explained the symptoms, and asked for a diagnosis, I'd probably have suggested carb ice right away. But in the heat of battle, solutions don't always come that easily. For these reasons, it's a good idea, in good conditions, to try to prime yourself for the worst. Of course, pilots might, as I did, overlook some potential complications. But even though I wasn't sure what my problem was, I knew what steps I would take if I couldn't come up with the solution. Throughout the flight, I had kept close watch on my position, I knew which direction I would turn for the nearest airport, and I knew pretty much how far I would let things go before I would call for vectors to get there. Although I may have been stymied, I was still in control. I had tried to plan for the uncertainty during all of my dull flights in good VFR cruise conditions. And I was about 50 percent successful.

Cruise flight is also a time in which pilots can use their skills and cunning to outwit the winds, increasing their speed and efficiency. No matter how well I prepare for a flight, I generally find that winds are not the same as forecast. So calculating winds for the altitude at which I'm flying is generally the first step in determining whether I can be doing better, in terms of speed at least, at another altitude. I often find calculating winds tricky business, especially without DME. So many variables enter into the equation that it's difficult to get an accurate idea of the wind's direction and speed if I'm off by a few knots on my own groundspeed figures, or if I stray a bit on my heading. If I find that I've got a good tailwind (almost never), I'll generally ask for a clearance to climb to stronger winds. If I find that I've got a headwind, then I generally think about descending. I have my conditions, though. I will always sacrifice a few knots for smooth air, and the weather conditions that I climb or descend into will have to be manageable.

Some pilots I know think that going through all of the calculations and gyrations to figure winds just isn't worth the trouble. They believe that by the time you've sacrificed the fuel and speed for a climb to high altitudes, you've given up the few knots gained in the altitude switch. And there is no guarantee that the wind vector at a higher altitude is the same, so the unwary pilot may climb, only to find himself with a less direct tailwind. Considerations like these are just part of the task of making the winds work for you. Sometimes they do, and sometimes they don't.

One of the things I've learned since I got my instrument rating is that the cruise portion of an IFR flight, the time when a pilot is working in the system, entails much more than my instructors ever let on. I always thought it a time to take a rest, or to start thinking about the approach. What I have learned from experience, and from other pilots, is that cruise can be one of the most important elements of a flight. In poor weather, interface between the pilot and ATC is crucial. Keeping your mind active and avoiding the great temptation to just "kick back" takes real discipline. Working in the system is one of the subtleties of instrument flying that was downplayed during my training, but I know now that it can go far in making a flight more comfortable, faster, and a good deal more interesting.

SECOND PART   Richard L. Collins

When we fly IFR we actually operate in two systems — air traffic control and weather. In training we tend to learn the two separately (if at all), but it is the interface between the two that makes the airplane a safe and reliable means of transportation. And if it were humanly possible to memorize all the minutiae in the *J-Aid* or *Airman's Information Manual* we'd still be lacking in ability to work the airplane in the two systems. It's just not in the books. It is something that is learned by doing. And remember when there is a conflict between a weather system and the air traffic control system, it's possible to get a special dispensation from the controller to deal only with the latter.

The en route portion of a flight has far more flexibility than the departure or approach. En route, once you decide the weather won't let you stay on the airway, the controller can allow you to zig, zag, detour, divert, climb, descend, and change destinations. Knowing when to choose one of these options is the primary technique.

Some general aviation pilots never make peace with en route IFR. They don't use the options and are on tenterhooks all the way — nervous, on the edge of the seat, convinced that some sinister force is going to "get" them. And well it might, if they don't learn to dodge at the correct time. These pilots never put together a mental picture of the possible and the impossible. There are very few of the latter situations, and the key to completing a maximum number of flights with a minimum of excitement is in identifying the impossible and avoiding it by using the flexibility of the airplane.

There are many en route considerations, but weather is foremost, so it will be discussed first.

BASICS

An understanding of the fundamentals of meteorology is necessary; it leads to sound weather decisions that are based on fact rather than guesses. En route, what we know and what we see is the basis for success. Remember that mistakes can be serious when dealing with weather; when it's really grungy out, every decision needs constant review. This is easy enough if we engage in a continuous weather refresher, using every flight as a lesson.

Even an apparent phenomenon like the thunderstorm varies from one encounter to the next — because it is affected by a lot of different factors. I studied this on a westbound trip one day and an eastbound the next, in the vicinity of Saint Louis.

When we were westbound, the controller told of an area of weather ahead. Other airplanes had been going around it, and the detours were long. I told Bradley, who was flying, to carry on. We'd go closer and take a look. I think he thought I was crazy.

Why head for what the controller described as a thunderstorm? The question was underlined by the fact that the Stormscope was showing a lot of electrical activity ahead. Only the weather radar in the airplane was on my side — it wasn't showing much precipitation.

My decision to continue was based on the weather pattern. There was a warm front to the west. We initially had flown at 14,000 feet but had found a very strong southwesterly flow up there. Lower, at 4,000 feet, we actually had a little tailwind on a westerly heading. A southwest wind high and easterly low is the classic pattern found to the north of a warm front. At 4,000 feet we were below all cloud bases and the visibility wasn't bad. I thought that the base of the thunderstorm activity was probably rather high — on the slope of the warm front — and that at our lower altitude we might be able to pass beneath everything. The strong temperature inversion we had seen on the descent, at about 7,000 feet, suggested that the feed of any activity and the corresponding updraft and downdraft activity would be weaker at our level.

As we drew closer, the Stormscope still said no — there was a lot of electrical activity. But the radar looked good, and we could actually see sunshine on the other side of the line of weather. There were breaks in the precipitation. The cloud bases appeared to be about 8,000 feet, and there were none of the telltale wisps of cloud that often define wind shear areas around thunderstorms. Clearly it would be okay to fly under at 4,000 feet. Equally clearly, if we had been higher we'd have had to go around the whole mess. The air traffic controller was cooperative and was surprised that we flew through the area.

The next day, eastbound at Flight Level 210, we came on the same thing, only different because we were at altitude. There was still a warm front shown west of the area, but a low pressure area was developing to the south. Our route was north of a severe thunderstorm-watch area, and the activity was much more widespread than it had been the day before. The controller said it was 75 or 80 miles through the weather, with imbedded thunderstorms. As we approached, the tops ahead were well above our level. The radar showed nothing but light rain, and the Stormscope showed activity to the north and south but nothing straight ahead. Normal air commerce was being conducted through the area, and there had only been a few pilot reports of moderate turbulence. Given those factors, I decided to go on through, straight ahead, on the airway. There were some moderate jiggles and a lot of snow. Precipitation static that interrupted navigation signals was the only real bother, and it went away as soon as we passed through the area.

If the Stormscope had shown lightning, it would have been necessary to deviate around the area of electrical activity. We were flying above the slope of the warm front and would have been flying in clouds with sufficient convective activity to produce lightning. That is not a smart thing to do. The day before we had been well below the clouds, with a comfortable margin both from the ground and the cloud bases. High, we were above the slope of the warm front. There is a big difference.

It takes experience with airborne weather radar and/or a Stormscope to develop good operational techniques, but I'd add quickly that after flying with them for seven years I was still learning. As time ran, I was ever more impressed with the need to combine a touch of weather wisdom with electronic wizardry.

A FLASHING NIGHT

An eastbound flight at night provides a clear example of how a basic knowledge of meteorology must be applied to all the things that you see and hear to make good in-flight thunderstorm decisions.

The FSS briefer in Indianapolis said there was a solid line of thunderstorms from Buffalo down into West Virginia, tops to 50,000, wind gusts to 55 knots, hail to one inch, and all the other bad things. The line was 25 miles through, according to a convective sigmet. The Pittsburgh radar report positioned it just to the west of that station and also reported that the activity was decreasing in intensity. The time of the report was an hour before sunset.

It seemed prudent to have a look. The airplane's WX-10 Stormscope and weather radar were available to provide information once we were en route.

Lightning was visible ahead soon after takeoff. This activity started around Columbus, Ohio, with some isolated cells; the real stuff was over Pittsburgh. The air traffic controller called with word that we would have to accept a slightly different routing because all inbound aircraft were holding for improvement at Pittsburgh and FL 190 was saturated in the area. That was okay, because the new routing also happened to coincide with an end run around the north side of the activity around Columbus. The air traffic and weather systems agreed.

Past that first bit of weather, the problem at Pittsburgh became clearly defined. A lot of lightning was visible, and it appeared solid, both visually and on the Stormscope. Wouldn't you know it, the radar shot craps at about this time. I still felt that I had enough information to continue on toward the line.

The controller was trying to be as helpful as possible. He said that the line appeared solid, but he added that a controller working a sector to the south reported a few breaks down that way. I started to go for that, but decided to continue straight ahead because a break appeared to be developing, both from lightning we could see and from the display on the Stormscope. I told the controller that we would carry on, and that if it didn't look right we'd divert into Pittsburgh, to regroup. There was no question that we needed to descend; the temperature at FL 190 was $-5°C$, which would mean a load of ice in any precipitation. There was also that desire to fly through a convective area at a low rather than a high altitude. I know that research has shown turbulence to be equally enthusiastic at all levels above the base of a storm, but I still like it lower.

There were a lot of things going for a continuation of the flight that evening. The pattern on the Stormscope was doing a perfect split in front of us, indicating a path free of lightning. Visual observations contradicted this slightly, with some cloud-to-cloud bridging of the gap between the two storms, but this neither showed on the Stormscope nor was very frequent. Also, we were at 9,000 feet by this time and had visual contact with lights on the ground a substantial distance ahead even though we were flying in moderate rain. Carry on.

The ride through was good, with the only bumps encountered on the east side of the weather. These came when we flew through cumulus with lower bases, east of the storm system.

The easy passage this night was caused by a couple of things. One, the line of storms was not associated with strong frontal activity. It was along a stationary front, and its development had been greatly enhanced by the heating of the day. Once the sun was down, improvement was in the offing. Two, the makeup of the atmosphere was such that the storms were based at a relatively high altitude. The air was quite warm up to about 9,000 feet, with a lot of instability above that level and not much below it. So the bases were higher. I knew about the nature of the front in advance; the part about the bases was learned as we flew along. I hasten to add that if things had not appeared favorable both on the Stormscope and to the eye, there would have been no hesitation to divert to Pittsburgh.

That flight would have been manageable without any on-board weather avoidance gear, but it would have been a nervous passage. The controllers were approximate with their information, but the spot where the line could be negotiated visually would have been easy to find, given the high bases and the good pattern of lights on the ground from Pittsburgh to the east. The nervous moments would have come once through the weather, in the little patch of turbulence that we encountered. With the Stormscope, I knew that the bad convective activity was behind, and that we were fooling only with cumulus clouds. They might harass and bang us around, but we certainly were not flying into the mouth of a monster storm. Without knowing there was no electrical activity, it would have been natural to wonder if continuing was a great mistake. Never feel guilty about continuously questioning decisions, and never hesitate to land for contemplative thought about the weather along a proposed route. Just remember that decisions must be made before you are up to your ears in alligators.

THE CONTROLLER'S ROLE

The role of the air traffic controller in weather avoidance is different from what it was in the past, before the controller's strike. Many of the people we talk with have relatively limited experience interpreting the relationship between their weather displays and light airplanes. Most are perfectly willing to tell you what they see, but that's all. And a lot don't fully appreciate the dynamic nature of weather. I saw a good example of this in the southeastern U.S., on a flight from Lakeland, Florida, to Baton Rouge, Louisiana.

Thunderstorms had been bedeviling the area around Pensacola for days. There was no front, just a lot of moisture and instability. The result had been torrential rains in the area.

As I approached from the southeast, I had to first get one thing straight. There are myriad restricted, alert, and military operating areas in the Florida panhandle and southern Alabama. Before continuing and "picking my way through" as the controller suggested, I had to know that any deviations from course would be approved without regard to military areas. I was once bounced badly because a controller couldn't let me deviate as much as desired because of military traffic; since, I've used gross avoidance of such areas as a procedure if restrictions apply. This day the good ol' boy on the ground told me that whatever I wanted to do would be just fine with him. He added that an airplane had gone through just north of the airway an hour ago, at my altitude, and reported a fine ride.

An hour ago. I appreciated the bit of information, but in a thunderstorm an hour is an eternity. Cells build, dissipate, and move. Specific information is really good for perhaps ten minutes, maybe less. You have to strike your own private deal with the air ahead on an immediate basis.

The area appeared solid on the King KWX 56 radar, which does an excellent job of defining weather. There was level three (red) along almost all of the line that lay ahead. The controller offered no advice on what looked best to him, so I took up a heading toward the area with the shallowest gradient from green to yellow to red on the radar. And I watched the Stormscope very carefully. The pattern there suggested that I was doing the right thing. All the electrical activity was either to the right or the left of the nose.

As the picture developed, the best route appeared to lie 30 degrees to the right of the course for a while, then to cut 45 degrees left through the narrowest part of the activity. I did this, and passage was in heavy rain for a brief while with some wind-shear turbulence but no strong updrafts or downdrafts. It was a good ride.

Just as I emerged on the west side, I heard the pilot of a

Cherokee inquire about the weather. He was going the other way, headed toward the area that I had just negotiated, and the controller told him only that other airplanes had been making it through okay. I called with a pilot report that there was heavy rain and some turbulence and that I had deviated north of the course. I felt that the controller might be making a bit light of the area. Without weather avoidance gear, I would have looked at it only at the minimum en route altitude, and then I would have flown only where I could see ahead. It was not a strong system, but there were active cells embedded within — maybe not strong enough to break your wings as long as control was maintained but surely strong enough to make you wish that you were elsewhere.

FUNDAMENTALS

A few flights in thunderstorm areas don't constitute an education, but they can provide the basis for a refresher on the fundamentals of the relationship between airplane, on-board weather avoidance gear, air traffic control radar, and the basics of meteorology.

For many general aviation pilots, the only available realtime information is from ATC radar and what is visible. And while pilots flying with on-board weather avoidance gear have come a cropper in thunderstorm areas, most thunderstorm accidents involve airplanes that don't have radar or a Stormscope. There is logic to this. Consider that a lot of pilots actually rely on information from the ground to lead them through areas of thunderstorm activity. Then consider that the traffic radar used by controllers is of limited use in serious weather avoidance. And even when and if it is combined with an overlay of weather — part of the FAA's master equipment plan — one has to combine information from the ground with other factors to use it effectively.

What's wrong with the controller's current information?

To begin, it shows rainfall rate only. The letter *H* is displayed on his scope in areas where heavy rain is falling. All a controller can tell you is where it is raining heavily. If he watched the weather return on the scope very carefully, he might be able to discern a pattern in the increase, decrease, or steadiness of the activity. But that is precious little information when it comes to judging convective activity. With airborne weather radar, it is the gradient — the distance over which rain increases in intensity — that counts. If you go from no rain to heavy rain in a very short distance, that indicates convective activity.

Another strike against ground radar is the fact that the controller's job is related to traffic, not meteorology. The controller who was telling me that the weather around Pittsburgh was all bad was considering only what he saw on his scope. He wasn't weighing the Pittsburgh radar weather report that the activity was diminishing, he probably wasn't considering the effect of sunset on the activity, and he probably didn't consider the synoptic situation and the vertical composition of the atmosphere. All he was considering were the *H* marks on his radarscope. So this time he made it sound worse than it was.

It works both ways. A controller might be looking at a scope that looks better than the one that night, yet conditions might be extremely unfavorable for passage. Pilots have found themselves in continuous wet and turbulent conditions in areas where traffic control radar didn't suggest anything other than scattered activity. This happens with developing low pressure systems or intensifying frontal systems. The time of day can have a lot to do with it, too. Surface heating during the day is a natural producer of instability.

I think we all recognize that weather forecasting is approximate, and I hope we recognize that the weather briefing for a flight isn't concluded until the flight is over. Many times we must piece together the real picture from information gathered before takeoff and en route, and from what we see and feel.

There are a lot of clues to developing thunderstorm activity and to the conditions that are conducive to development.

A southerly wind that is stronger than forecast is a signal that storms might develop more rapidly or be stronger than expected. Why? Because in most parts of the United States a flow from a southerly direction bears moisture, the stuff of which thunderstorms are made.

We don't worry a lot about ice in the summertime but the forecast of temperature aloft is worth jotting down during a weather briefing. If the actual temperature aloft is colder than forecast, that means the air is more unstable than expected. Instability is one of the ingredients of a thunderstorm.

Cumulus development also tells a tale. If cumulus clouds are building rapidly, and if when you fly through them they are both wet and turbulent, chances are they are building to bigger and better things. If when flying toward a front or toward the quadrant northeast of a low pressure center you encounter taller, wetter, and bumpier cumulus as you go along, things will get worse before they get better. The basics of surface circulation tell us a lot about where those fronts and lows are located. The wind flow is counterclockwise and inward around a low, modified a bit by fronts. It's usually southeasterly or easterly north of a warm front, southwesterly south of a warm front and east of a cold front, westerly or northwesterly west of a cold front, or northerly or northeasterly northwest of a cold front that is becoming stationary.

THE SIGMET SYSTEM

There's never a real excuse for an airplane to get into trouble because of severe thunderstorms. For one thing, these are widely advertised, if not by the government at least by their presence.

The system of convective sigmets works well. These cover existing thunderstorms and areas where development appears likely. Sigmets are as close as we come to realtime thunderstorm information, short of looking at a radar set. We can't use a sigmet as anything other than one bit of information, though. There have been airplanes lost to thunderstorms in areas where no sigmet existed — the Pan Am 727 at New Orleans is a prime example — so the lack of one shouldn't be taken as any guarantee weather is not there. Remember, the sigmet is a response to a situation, and while the response is rapid it is not instantaneous.

Once the sigmet exists, it must be used as a definition of a problem that we must solve. The sigmet paints with a relatively broad brush. An airport within an area covered by a sigmet may or may not be usable. A route may or may not be flyable. I like to think of sigmets as warnings. And I must say that a pilot who operates with open eyes and ears and who has some weather avoidance gear on board will never be surprised about the issuance of a sigmet covering any part of the route ahead. The left seat of the airplane is usually the best place to learn about storm activity.

Springtime is often thought of as thunderstorm season, but on one fall trip I was exposed to the same stormy weather system several times. It had been a dry summer, and the trip served as a good thunderstorm refresher.

HOT SPRINGS–SAN ANTONIO

The first encounter was on a leg flown from Hot Springs, Arkansas, to San Antonio. I knew there was some activity down that way, but the weather map I had drawn showed no spawning feature, so I just naturally assumed that any storms would be of the scattered, air mass variety. Soon after takeoff at Hot Springs I got the first clue that the storms were more than scattered. There was widespread talk of deviations on the center frequency. Also, the Stormscope in my airplane displayed substantial electrical disturbance ahead. The radar wasn't showing it yet, so I knew that the storms were more than 80 miles ahead, that being the effective range of even a good radar with a ten-inch antenna.

Soon, the storms were shown on radar. The first ones appeared to be rather mild, but certainly strong enough to warrant an alteration of course. Then they got livelier, but I convinced myself that I could see the end of what had become a line. I thought I could avoid the activity by steering right, over Waco, Texas. This was where adding one more bit of information helped.

The center controller said I could go as I wished, but that the big picture on his radar suggested a detour all the way around by Lampasas, west of Waco. Then it should be possible to cut south into San Antonio. One reason for his suggestion was the restricted airspace south of Waco. He wasn't interested in arguing with pilots about deviations around weather and into the restricted area. The weather return on his scope must have intensified after I talked with him because he virtually insisted that a following airplane make the trip around to the west.

As I passed Waco, I turned momentarily to the south to have a look with the radar. It didn't look bad, and the Storm-scope wasn't exactly lighting up in that direction. I started to see if I could go direct to San Antonio, but stuck with the controller's plan.

That he was correct was shown as I neared San Antonio. There was a huge mass of thunderstorms over town and air-port, extending to the northeast and east. I'd have had to fly through that coming over the filed route; from the way I came, I only needed landing weather, since the airport was on the western edge of the almost stationary system. It was strong enough to dump four inches of rain on San Antonio, more rain than they had had in a long time. Clearly, even with all the available on-board equipment in this case, I had gotten my most valuable information from the air traffic controller.

LITTLE ROCK–DAYTONA BEACH

The next leg on which I encountered this weather system was from Little Rock to Daytona Beach. (The trip from San Antonio to Little Rock occurred early in the day, before much activity built up.)

In my briefing, the flight service specialist was uncommonly optimistic. He said that the coverage was 10 percent thun-derstorms, 40 percent light rain to the east. There was an active cold front approaching from the west, and all the bad stuff was out that way. He opined that I'd probably be on top of almost everything at Flight Level 190. Maximum tops were given as 26,000, and he saw that as an isolated buildup.

PIECE OF CAKE?

I prepared for the flight with the thought that it would be a nice and easy ride. But when I looked my radar squarely in the eye after turning on course, I decided that it would not be a nice flight. There was a lot of rain out there, plus some sparks on the Stormscope. As I climbed, I could see building cumulus in the murk. The air was lightly turbulent, and I started moving around the areas of strongest return on the radar, and around all areas of electrical activity as shown on the Stormscope. This day the air traffic controller couldn't be of much help. The center controller simply said that his scope was relatively full of weather for the next several hundred miles and that I was free to deviate around stuff as I saw fit.

Something else raised its head as I climbed. The temperature was dropping, and there was a lot of moisture. I decided that I had best level off below the freezing level. It was pretty much a maximum-range trip to Daytona, given very little tailwind, and I didn't need to lose any speed to ice. After I had been at 15,000 for a while, the controller gave me a message that made the decision seem a wise one. An air carrier flight reported moderate icing from 16,000 feet to 24,000 feet. I sure didn't need that.

My airplane was in cloud, and I was working hard at interpreting the radar and the Stormscope for a couple of hours out of Little Rock. Just after I cleared that area, the controller broadcast a convective sigmet for the territory I had just passed through. I quite agreed with the observation and reflected on how you had best be ready for every eventuality on every flight. The trip that I thought would be easy wasn't. There were both thunder and ice to think about.

I got another refresher on thunderstorms and restricted airspace when approaching Daytona Beach.

The usual Florida scattered thunderstorms were about, and I had one between my position and Daytona as I passed over Jacksonville. When quizzed, the controller said to deviate around to the west side of the storm. I asked again, because central Florida is full of military restricted areas and I didn't want to be squeezed between a storm and air not available for flying. The controller answered the second asking of the same question rather testily: "Yes, I say again, deviate to the west."

Then, right before I got to the storm, the controller called and said to turn left to a heading of 140. That just happened to correspond with the exact center of the thunderstorm I was avoiding. When I told him that I couldn't do that, he shot back with an instruction to turn to the east and said that I was about to enter a military restricted area the way I was flying. I tried to be patient in explaining that that was why I had asked the question twice about deviating to the west, and I suggested that he put the rock in his other hand. That drew no comment, and I flew the extra miles of a westerly deviation turned easterly with resignation. This was but a couple of years after the old controllers went on strike. There was no way to expect perfection in folks with less experience. I got another lesson on this before landing. The cheerful female controller at Daytona told me that I was around the weather and could fly direct to the airport. But there was a boiling cu there between my little airplane and the airport, topping at probably 15,000 feet. When I told her that I would be flying another few miles to the east, she sounded puzzled. "Why, it looks fine to me for you to fly direct."

"Okay, ma'am, if you want to fly a light airplane through clouds that look like that, it's okay with me. But I am going around them."

SQUALL LINE

My next encounter with this piece of weather came as I flew from Daytona back to my home base in Trenton, New Jersey. By this time an active cold front had developed. I'd been just ahead of the front on my flight northward. And, as often happens, there would be a line of thunderstorms ahead of the cold front.

The line was oriented northeast to southwest and was about 20 miles west of Jacksonville. I had filed along the airway that runs 20 or so miles offshore, over the Atlantic. There is restricted airspace over the ocean just east of that airway. Additionally, I draw the line on overwater flying in an airplane that isn't equipped with flotation gear. I once flew an airplane with an operating limitation that read as follows: "Intentional Ditching Prohibited." I heartily agree and only fly over large bodies of water with enough altitude to glide to shore.

My problem this day was twofold. One, I had to determine that my offshore route would be flyable virtually without deviations and, two, I had to maintain an option. In this case it had to be flyable weather behind. To be forced out to sea by a squall line would be one of the most unintelligent things that a pilot could ever do. Truly you'd then be between the devil and the deep blue sea.

In examining the return on my radar and on the Storm-scope, I thought that I'd just clear the north edge of the line in flying the airway. But I didn't really trust this information. When flying at such a shallow angle to a line of weather, it's easy to engage in self-deception about the end of the line. Certainly the radar picture is affected by attenuation (meaning that cells behind other cells wouldn't show because the nearer return is reflecting virtually all the energy emitted by the radar), and certainly I wouldn't want to bet everything on the Storm-scope being accurate within five degrees when viewing a lot of activity from a distance. Once again I had to rely on the controller. I asked him to check with the next controller and verify that the line would indeed allow passage along the air-way. I also asked how fast it was moving. The very helpful lad got the information for me and said that based on my progress and the line's eastward movement, I would be able to clear it with only a slight deviation to the east.

It worked fine, but I wasn't kidding myself about some increase in risk. Sure, I was within gliding distance of land. But for about 30 or 40 miles of the flight there was a squall line between my airplane and that land that I would glide to in case of trouble. The risk was ever so slight, but I like to ac-knowledge each flirtation with fate so that I can reflect on it later. The question always has to be, "Was it worth that little extra risk to complete the flight as planned?"

There was still a lot of scattered activity north of Charleston, South Carolina, but it was possible to pick my way around it using the radar and the Stormscope.

I had strong thoughts about weather avoidance equipment as I flew along. I avoided thunderstorms for years without the gear, really trespassing only once. I'll always remember that wet and bumpy ride and have no desire for a repeat. In flying at lower altitudes than the middle levels used in my pressurized 210, I thought that I could do a good job of staying out of storms without the equipment, and with not too many canceled trips. But despite the feeling that flying higher means you'll be on top of most weather, I have learned that the high teens and the low twenties are actually the most difficult altitudes at which to deal with thunderstorms. You are in the clouds a lot of the time, it's more difficult to make a good evaluation of what you see, and there is a lot of ice at those levels in the warmer months. If I didn't have both a radar and a Stormscope I think I would fly below 10,000 feet in the airplane anytime that storms are on the prowl. With both those items, I'm quite happy at the higher levels. The only thing that would be better would be a bigger antenna for the radar, but that's not really viable on a single-engine airplane.

I thought that the end of this flight would be pretty easy, based on the forecast. They were calling for 4,000 scattered, 8,000 broken, with occasional lower in light rainshowers and the chance of a thunderstorm at my destination. That sounds easy enough.

The radar was full of rain as I flew into the Philadelphia terminal area to begin my approach to Trenton. There were even some sparks on the Stormscope, and in the low light of the end of the day I saw some weak lightning. Some aircraft were requesting deviations around weather, but I couldn't honestly see where one direction was any better than another.

One thing that I failed to do before starting my approach was refresh myself on the conditions that I'd be flying through to land. I had the ATIS — it was 800 broken, 3,000 overcast, and 2 miles visibility — worse than forecast. And looking down, the low clouds looked as if they were closer than 800 feet to the ground. They were also moving very smartly out of the south. The wind at 3,000 seemed southwesterly to westerly and not so strong. The surface wind at Trenton was given as southerly at 15 knots.

I still didn't think of it as a challenging approach, and even asked the tower if I could have the option of landing straight in on Runway 6 as opposed to circling for 16.

That was purely foolish. At about 2,000 feet on the approach I started encountering very strong wind-shear turbulence. I had no excuse for being surprised, and when I saw the airspeed running 30 knots higher than is normal for the power setting and configuration, it was something that I could and should have anticipated. That's what comes with a rapidly decreasing tailwind as you descend. This and increasing drift to the left scotched any idea I had about making a straight-in approach. The rain was very hard, and I wasn't convinced that I'd be able to keep the airplane on a wet runway with that much cross-downwind — to say nothing of stopping.

It was almost dark by this time. The most difficult maneuver of all is a nighttime circling approach in wet and bumpy conditions, and I was just about to do this very thing. To boot, I was barely clear of clouds at the circling minimum. Here I gave myself a mental kick in the pants and suggested that I get on with doing this in a professional manner by making the best of the situation. To me that meant flying up over the airport and circling left instead of breaking off and making a right base to Runway 16. I felt by overflying I could better size up the pattern. It all worked well, and I soon was getting soaked while winching the airplane into its hangar. Then, as I closed the Jeppesen book before putting it away, I decided to award myself the Stupid Prize for the day. There is a straight-in RNAV approach to Runway 16, and I hadn't even thought about it. I could have approached straight-in as opposed to circling, and I didn't even ask. That is a clear case of not considering all the options. Bad grades were deserved for this.

Examples can be used to outline how we have to roll with the punches and accept the weather as it presents itself. An incorrect forecast is no excuse for a problem. There is really no "surprise" weather. It is there, as real as the sky itself. The only pilot who is taken aback by the weather is the one who ignores the obvious.

BUSYWORK

On a majority of our flights the weather is actually rather simple. If there is such a thing as an average flight, it probably involves only a few minutes of flying in clouds. The rest of the time, the airplane is in a most benign situation: smooth air that is clear of clouds and a relatively simple and routine ride. But it isn't a time to let the brain coast. Bradley is sure right when he extols the virtues of keeping busy.

En route fuel planning is a good example of necessary work that we sometimes fail to do. If there is a fuel problem it starts long before the actual act of running low on (or out of) fuel. I've never run out and don't intend to, but I can certainly imagine how it all starts. All of us like to avoid extra fuel stops. And any person is capable of engaging in self-deception. But just as I can look anyone in the eye and say that I have flown an airplane over its gross weight only once in the past twenty-five years (the line crew exceeded my fuel order, and there was no way to get the fuel out — no excuse but that was the reason), I can also truthfully say that I haven't landed with less than enough fuel to fly an additional forty-five minutes, and that was fifteen less than the plan.

Perhaps some would think it a fetish, but on a maximum-range flight I continually recalculate my fuel reserve. I might fill up a whole piece of paper with numbers in the process of doing this. And any time the interface between the fuel remaining and the weather ahead becomes questionable, I stop and buy gas.

We all learned the rule about alternate airports when getting an instrument rating. This rule is okay, but it is no substitute for common sense. A forecast may indicate that a place will be good enough as an alternate and the place might fold flat on its face — zero-zero. By the same token, a place for which the forecast is bad might turn out to be the most viable alternate on a flight. But no alternate is any good if you lack the fuel to get there.

If you fly exactly according to the rules the process can become a charade. It simply doesn't make sense to go through the business of flying to a destination, finding it below minimums, holding for forty-five minutes, and then landing at an alternate and running out of fuel taxiing in. The real goal should be to land every time — destination or alternate — with an hour's worth of fuel in the tanks. That's really quite easy; the alternate, however, will seldom be the one on the flight plan.

For example, let's say that you are flying from New Jersey to Indianapolis, Indiana. For some reason, when the weather is cruddy out that way, Cincinnati, I have found, is most likely to be the place with legal alternate minimums in its forecast. Nothing else in range ever seems to work. And many times I have figured the fuel down to the last drop, using Cincinnati as an alternate. (I fly this leg at least once a month and have for years.) But I have never used Cincinnati as an alternate. The few times that Indianapolis has been below minimums,

Dayton (about 80 miles this side of Indianapolis) has had minimums, and I have landed there and topped the tanks. I just wouldn't press on to Indianapolis (or anywhere else) with the weather reported below minimums unless there is a lot of fuel in the tanks. The time to make the low-fuel decision is when it first appears that the stuff will be short. Even the most mediocre pilot knows early-on when a flight is going to leave fuel in short supply.

WIND

The wind aloft is a major part of the fuel equation and demands a lot of en route understanding. The first thing to understand is that the winds-aloft forecasts are a rough estimate, seldom accurate and often so inaccurate that preflight planning becomes relatively worthless. In some instances the wind forecasts don't even reflect the general flow aloft. Like weather, wind is what is there, not what is forecast. We fly in it as it exists, and running short of fuel because a wind forecast is wrong isn't excusable.

There are some general ways to anticipate wind. One big item relates to the fact that wind speed increases in a frontal zone. Because forecasts are general and cover a relatively long period of time, they don't reflect this. If you are flying toward a cold front with a 30-knot headwind, you can bet that the headwind will increase. After passing the front, the wind will likely decrease but that takes a while. On a longer flight it is highly unlikely that the wind will remain the same regardless of what might be forecast. The variations in wind are why fuel planning must be done on a continuous basis. I've found a high degree of inaccuracy in estimating times of arrival based on the groundspeed at the beginning of the cruise portion of a flight.

Also, it can't be stressed too often that a substantial error in the winds-aloft forecast means that all other forecasts are suspect.

One fall evening, I was flying from Wisconsin back to New Jersey. The weather along the route was perfect, but there was a tropical storm off the Carolina coast. The storm was not forecast to affect the weather in the northeast. All the forecasters were predicting lovely warm days. But I found the wind aloft to be both more easterly and much stronger than forecast. I decided that any forecasts of nice weather were suspect. My suspicions were confirmed the next day as I sloshed down 32nd Street in New York to work. Wind aloft or on the surface tells us a lot about the weather; when forecasts don't get either right then you know that the computer's idea of weather patterns is basically incorrect. The same is true of temperature-aloft forecasts. If these are wrong, then the computer's idea of the nature of the air is incorrect. There is a direct relationship between temperature and the air's ability to hold moisture (warm air holds more); temperature as well is an indication of the stability of the atmosphere. So this is rather important.

EVERYONE A LERT

Be alert, the world needs more lerts. At least that's what the kids say. It has special meaning to the IFR pilot because instrument flying is a heads-down activity but we mix with VFR traffic when flying in visual meteorological conditions below

18,000 feet. There's no way the controller can or will call all the VFR traffic, either. Any time we aren't alert for other traffic, we aren't doing a very good job of flying. A very special place to be doubly alert is when flying between 10,000 and 18,000 feet. There is no speed limit in those altitudes, and some business jets take advantage of VFR operation on short trips. The closure rates become very high, and if you aren't working at the business of searching the sky, you'll not likely see one until it has passed. There is a requirement for a transponder and encoder above 12,500 feet, but a controller still might not call traffic.

THE OLD BOD AND MIND

While fatigue becomes more hazardous on an approach, it is generated en route. There's no way to avoid fatigue, but there are ways to minimize it. One is by flying in smooth air. There are times when we are tempted to fly low, in turbulence, to minimize the effects of a headwind. This is purely foolish on a long trip because of the fatigue factor. To lurch along for three hours is a lot more tiring than moving smoothly through the sky for three and a half. I'll always opt for smooth air, and find that it is almost always available en route. Oh, there might be bumpy moments or even bumpy spells, but I can't remember when I have flown an entire leg of any length in continuous turbulence. This is easier to do in a turbocharged airplane, but even in a normally aspirated airplane it's usually possible to find smooth air except over the high country in hot and windy weather.

Another way to minimize fatigue is to stay busy. Being bored is probably one of the most tiring things in the world, so if you sit and stare at the clock, you'll be more tired for that approach than if you stay busy with fuel planning, route review, and contemplation of all the possibilities. The latter includes the time-honored game of "what would I do if . . ." If the engine fails, if the alternator fails, if the vacuum pump fails, if smoke starts entering the cockpit, if the instruments become unreliable — the list is endless.

En route flying is really what we make of it. In general-aviation IFR, there are a lot of en route accidents. In airline flying there are almost none. Some might argue that this relates to airline equipment and two- or three-person crews. But I think it relates more to their continuing proficiency maintenance programs and more attention to the subject. Sure, airline pilots ride along and talk about things other than flying. But they are all business most of the time — especially when it counts. Also, while they do get paid to take people to the desired destination, they also get paid if they cull that original destination and go elsewhere. We need to think of our flying the same way.

•

CHAPTER 5

Arriving

FIRST PART   Patrick E. Bradley

One of the easiest ways I've found to botch what could have been a good approach is to foul up the descent from cruise to final approach altitude. "Twenty-seven Q is cleared to descend to three thousand." The words seem simple, but I have fallen prey to the many complexities that a straightforward descent presents. For the pilot of a pressurized piston airplane cruising between FL180 and FL290, descents are most difficult. But even at lower altitudes, pilots of piston-engine airplanes can have serious difficulties combining the tasks of getting the airplane down in time and of preparing the airplane for the upcoming approach. Further complicating the procedure are the frequent changes — affecting altitudes, direction, speed, and even approaches — that undermine even the most complete planning.

I sometimes have problems timing my descent so that I reach the initial approach fix at the proper altitude. One reason for this is that I may begin a descent too late, or ATC might not give enough time for the airplane to safely descend. No matter how quickly a pilot wants to descend, he is limited by concerns of shock-cooling his engine and maintaining a safe airspeed. Both are definite limits on descent rate. The ultimate problem, and the one that most often catches me, is realizing that when the airplane isn't going to make the assigned altitude it's time to make new arrangements with ATC.

My worst descent experience came at Jackson, Tennessee. Cleared from FL 210 to 13,000 feet, I began the descent far enough out to reach the initial approach altitude by the time I reached the initial approach fix. Center had other plans, though, as is often the case. Reaching 13,000 feet, I anxiously awaited clearance to continue the descent. As the miles ticked away, though, I still hadn't heard from center. I called and requested clearance to a lower altitude. But for reasons unknown to me, I was told to stay right where I was. By the time I was cleared to descend, I knew that it was going to be one of those approaches. Sweating out every foot, I inched downward. At lower altitudes, where the airspeed increases in proportion to groundspeed, I was forced to further decrease the descent rate to keep the airspeed in the green. I just couldn't get the airplane down in time to make the approach.

My most serious mistake during that descent was not taking affirmative steps when I realized that it was not going to work. Long past the point at which I should have asked for a hold or some vectors to continue my descent, I was still struggling madly, and futilely, to get the airplane to descend faster than was possible, and faster than I could think. From that experience I learned that when a descent and an initial approach fix aren't going to come together, it is time to take advantage of other options. ATC might get ruffled, but they won't refuse to help out in such a situation.

Part of the reason I neglected to make a new plan during my descent was the weather. Ceilings at Jackson were about 700 feet, and through much of the descent, I was in some pretty bumpy clouds. My concentration was focused so intently on getting the airplane down and preparing for the approach, I just didn't consider the eventuality that I would scream through the initial and final approach fix way too high and way too fast. It took some assertive guidance from the right seat to redirect my concentration. I wasn't thinking far enough ahead to know that without two steps backward and a more ordered descent, I'd just end up going around anyway. While I was trying to get set for the approach, I was oblivious to the fact that I was never going to make it.

One way I've found to avoid, or at least predict, many of the pitfalls that may arise during an approach is to think it through before actually starting down. Knowing the weather you will be descending into is always a big help; in some situations, it might be preferable to chose an alternate before even beginning the descent. For instance, if icing is a potential problem, it might be wise to make a steeper descent rather than a long flat descent. That way, the airplane spends less time in icing conditions. On the other hand, if ice is going to present problems and the weather at the destination airport is right down to minimums, it might be worth finding another destination with more favorable conditions. Even if there is no ice but five airplanes have missed the approach in the last half-hour, it might not be worth attempting the approach. Deciding to miss an approach before beginning the descent can save fuel and open new options. Each of these scenarios requires a judgment call, but weighing the options before beginning the approach seems a lot simpler.

I also find that it helps to visualize, to the extent possible, where my descent path will take me in relation to the airport and significant terrain. Often, ATC will change your proposed path by issuing vectors, but when you know your relationship to the airport and terrain from the starting point, it's easier to account for changes. One of my instructors told me the story of a pilot he heard speaking to ATC. After being cleared for the approach, the pilot asked the controller, "I'm a bit disoriented, here, could you tell me where I am?" That pilot's admission obviously struck a chord in my instructor. He grilled me on positional awareness before nearly every approach. He would give me headings throughout the descent and expected me, at any given moment, to be able to give my location relative to the airport. When the airport had an NDB or locator beacons, the task was generally pretty simple. But when I had to determine my position on VOR radials or my position relative to the localizer, the task became much trickier. I found that I had to keep track of the vectors I was given and follow my direction in relation to the approach plate diagram. If I stayed on my toes, it was simple to keep track of my position. If I woodenly followed directions, it became quite difficult, on the spur of the moment, to figure out where I was. At other times, my instructor would just pull the throttle back, announce that I had lost my engine and was out of range of the airport. My hypothetical ceilings were generally 200 feet overcast. "Where are you going to go?" my instructor would ask. I would make my best guess, knowing he had probably given me the choice of open fields or a 1,000-foot ridge. When I turned toward the ridge, he would let me descend to about 500 feet over it and then have me take off my hood. His point was well taken.

Besides reviewing my descent course, I try to check two other points before I begin a descent. First, I make sure that I know the initial approach fix and the altitude to which I can descend when I finally do get on a published segment of the approach. Before that time, I know that I should be on an airway at or above the minimum en route altitude or above the minimum safe altitude, when within 25 miles, as published on the approach plate. I've never had a controller clear me below the minimum safe altitude before I'm on a published segment of the approach, but this way I've got a double check that will prevent any misunderstandings. Second, I try to run through my mind the points at which I will go through the various checklists. I am pretty good at following checklists, but I am very bad at remembering to perform them. When I can associate a checklist with a particular point in the descent, I am more apt to remember it. For instance, if I am cleared from a middle cruising altitude to the minimum safe altitude, I usually chose that time to switch tanks if necessary, and to start setting as many approach frequencies as I can. At this point, I also run through the memory items for the approach. Many pilots wait until later in the descent to begin preparations for the approach — for instance, a few thousand feet above the airport. But I find it simpler to link checklist items to an event rather than to an arbitrary number, if I can. My primary goal is to stay a step ahead of the airplane and to be ready, when I hit the initial approach fix, to concentrate on the approach and the final checklist items.

My biggest problem in the descent is my tendency to look at it as simply a time to descend from cruise altitude to approach altitude. I guess that this is why it is most often in the descent that I set the stage for a rushed, harum-scarum approach. I tend to put critical checklist items off to the last minute, and when the time comes, I end up scrambling to complete them or forget them entirely. I do best when I look at the descent as a transition into the approach and mark definite steps to definite points along the descent. What makes checklists so complicated are the descent timing and position checks that vie for the same limited time. All of these considerations — preparation, descent timing and position — go hand in hand in a good descent, and I suppose that this is why, for me at least, the descent is one of the trickiest elements of an IFR flight.

SECOND PART   Richard L. Collins

In the training process, pilots don't get much exposure to arriving. Instead, they practice approaches and holding patterns over and over and over. Then, when the real world shows up in the windshield, problems develop. Just as Patrick didn't know what to do with a lot of extra altitude at Jackson, Tennessee, a lot of pilots find the link between cruising and shooting the approach to be elusive. There is a lot more to the arrival than the approach, and an actual complete arrival often bears little similarity to the approach depicted on the chart. There is a lot to do in an ever-shortening time.

The arrival problem was illustrated to me twice in a brief period. Two pilots were coming to visit. Both had relatively new instrument ratings. The weather was close to minimums for both. And both had trouble on the approach. The problems encountered were almost identical. They were vectored to intercept the ILS just outside the marker. They both reached that point going a bit fast and with too much altitude. The approach was starting, but the pilot was not ready. When the controller told each that he was a mile outside the marker, cleared for the approach, call the tower, the pilot suffered a mental overload. There was too much to do. Both pilots completed the arrival — one after a missed approach and another try, the other after a salvage job that left him a little shaky. In fact, instead of returning to base that evening he opted to spend the night.

Such events are common and happen to even the most experienced pilots. To counter the arrival that goes to pot because events don't unfold as we think they might or because things happen rapidly, we have to refresh our ability to think and plan during the time leading up to the approach. Actually, it is a lack of thought and planning that allows an airplane to get ahead of a pilot on an arrival. The importance of all this is underscored by the fact that a high percentage of the IFR accidents happen on approaches, and that poor arrivals tend to lead to poor approaches.

For the purposes of this discussion, arrivals start with the clearance to start descending from cruise altitude, and they evolve into approaches at the final approach fix, when you are inbound to the airport.

As the airplane gets closer to the destination, positional awareness becomes increasingly important. "Here I am, at ten thousand feet, X minutes from the airport." X, the time to the airport, is a key part of the equation. If the descent is started too late and becomes a hectic time, with some doubt about getting down before getting to the final approach fix, then the groundwork is laid for further problems. More things have to be compressed into less time. Ideally, we plan to have about five miles of straight and level flying before reaching the final approach fix. At some airports this can't be done because of terrain or local procedures, but it is possible at most places. The goal is to do it; real skill is shown by handling the situation when it seems impossible.

I use a rather simple formula to determine where to start a descent. In piston-powered airplanes, I start the descent 5 miles away for each 1,000 feet to lose. That is based on descending 500 feet per minute at 150 knots. For each 30-knot increase in speed, increase the rate of descent by 100 feet per minute. (For a 30-knot decrease, cut it by 100 feet per minute.) That doesn't mean that I always descend at 500 feet per minute. I just use this as a guideline. For example, if the air is turbulent or hot down low and I want to stay high longer, I use this as a benchmark distance to tell me where about 500 feet per minute would have been required. If I wait until later to start descending, I know that the required rate will be higher and the time shorter. How much higher and shorter is easy to figure on a calculator or in your head if you are one of the few humans who can still multiply and divide.

ENGINES AND EARS

You can hurt two things in a descent that is too rapid. If the power is reduced drastically, the result is rapid engine cooling. They call it "shock cooling," and some of the first pressurized piston airplanes had trouble in this regard. With pressurization, the pilots were not worried about hurting in the other area — the ears — and they just honked back on the power and came right on down. The result was very high engine maintenance bills. Better to plan ahead and come down gradually.

DESCENT WEATHER

If the trip is fairly long, we get used to our cruise weather. Often it is smooth and in the clear. It's sure not going to get any better in the descent, and it might get worse. Considering the expected conditions for the descent has a direct relationship to planning.

It's good to estimate the level at which the air will become turbulent as well as the level of the turbulence. If there is some possibility that it'll be bumpy enough to warrant slowing to the turbulent air penetration speed, then either the rate of descent will have to be less, the engine will have to be shock cooled in the descent, or you'll have to use whatever drag-producing device is available — which almost always means the landing gear. Maybe if it looks like it's going to be smooth to 5,000 feet and rough from there on down, it's a good idea to descend as rapidly as practical to 5,500 and then reduce the rate of descent and the airspeed before entering the area of turbulence. (Personally, I never use airspeeds in the yellow arc, even in smooth air. For me the top of the green is the limiting speed.)

When you are flying very high — above 10,000 feet — there is a good chance that the conditions during descent will be entirely different from conditions at cruise. Is there any special consideration in the descent-area weather?

Ice might be one factor. In the wintertime, the lower levels can be pretty icy, as can the tops of stratocumulus clouds. Descent planning means minimizing the exposure to ice whether or not the airplane is equipped with deicing gear.

Consider a day when there are stratocumulus clouds about. A typical situation might be behind a cold front that is becoming stationary. The stratocu tops are often between 8,000 and 10,000 feet; the air at the surface is usually cold. In a situation like this, you'd have most likely been cruising on top. It would be best to descend as rapidly as possible through the top couple of thousand feet of cloud. At least it has always seemed to me that the most ice is in the top couple of thousand feet of stratocu, and it can be present at very low temperatures. I recall one day when it was −15°C on the descent (and even colder on the surface). I started getting ice on a Cessna 402 in the tops of the stratocu and couldn't get an altitude change right away. I droned on, with the prop anti-ice operating, thinking that would keep the blades clear. After about a quarter of an inch of ice accumulated on the wings, I cycled the boots. It didn't do a lot of good, and within a few minutes I started feeling the slight vibration that comes from ice accumulating inboard of the electrically heated boots on the prop. Fortunately I was soon out of the condition and on the ground. We had a tough time getting the ice off the airplane before flying it again because of the extreme cold temperatures. Had I planned the descent better, the exposure would have been minimized. I should have told the controller that I didn't want to start down until I could have an unrestricted descent to a relatively low altitude.

Freezing rain is a condition that requires very good descent management, if not a trip to an airport where the condition does not exist.

Take as an example a condition where the temperature is below freezing within 3,000 feet of the ground, with warmer air above. A typical approach might call for the airplane to be vectored for the approach at an altitude 2,000 feet above the ground. But that wouldn't work. The only way to arrive in freezing rain is with an agreement that the airplane will remain in warm air until an unrestricted descent can be made. Even then the risk is high. Ice can accumulate rapidly in freezing rain, and you might find yourself landing a very sick airplane with no forward visibility.

Icing conditions take many forms; the main thing is to consider all the possibilities during an arrival. And just because you are thinking in relation to a descent where you want to come down anyway, don't give ice short shrift. It might arrange for you to reach the ground before reaching the runway.

If you fly at levels above 20,000 feet, you might be on top of convective activity at cruise and will have to descend through it on arrival. That's something to think about in advance and plan for. If the tops are below 20,000, the weather usually isn't severe, but it can be turbulent enough to require a more gradual descent than you might normally like. That means starting down farther away, or perhaps descending with the landing gear extended, in a retractable.

DESCENT CHECKLISTS

For some reason, those of us who fly single-pilot are less likely to use a checklist than a crew of two. This is especially true in flight. For example, we seldom whip out a descent checklist to make certain all items are handled. That doesn't mean that it's not a good idea to have a checklist for everything and to use it. On descent, there isn't a lot to remember, but if we follow the good practice of setting as many things as possible as far in advance as possible, the approach itself will be easier. Fuel, lights, whatever might be required that can be done ahead of time — they all *should* be done ahead of time.

I use certain points in the arrival as triggers for a checklist. Descending through 10,000 feet or leaving a cruising altitude at or below that level is a point where I run this list:

Fuel   I like to select a tank that has enough in it for the rest of the arrival, the approach, and the landing or the missed approach. I think it is best to make any final fuel selection at this point than to do it later. There is always the possibility of an error when switching tanks, and I'd rather cover that error higher than lower. It also gets the action out of the way, leaving only a verification of the fuel status for the before-landing check.

Lights   If descending in instrument conditions, a lot of lights are not necessary. But if it's VFR down below, best pop out of the clouds with as much lighting as possible. Lights are very effective on cloudy days so there's real value there. No strobes in clouds at night, though; they'll make you dizzy.

Landing information   If the airport has an automatic terminal information service, this is the time I listen. Some choose the time of the handoff from the en route controller to the terminal controller, but I like to have the straight scoop earlier if possible. Getting the information prompts a review of the approach plate, and there's more time to do this early in the descent than late. If there is no ATIS I get whatever information is available. In the case of an airport without weather reporting, I get the nearest weather and contemplate what differences there might be between that and the conditions at my destination.

Pilot   The arrival is a different time than cruise, and it is time to remind the pilot-in-command of a few things. At cruise a certain level of sloppiness is not critical. It certainly isn't good to accept this, but we'd all be kidding ourselves if we contended that pilots must fly as precisely when en route as when arriving. If you are a couple of miles off the airways while flying along, it's no big deal. The controller probably won't call and the passengers will never notice. But if you are a couple of miles off on landing, everyone will notice.

We have to go through the process of getting psyched up for the arrival, approach, and landing. I adjust my seat if it was changed for cruise, which it almost always is. If I don't have my headset with boom mike on, I put it on. Headsets drive me batty if worn for a long time, and as a result I shunned the use of them for a long time. But now I wear one when arriving and departing, because it makes things a little easier.

The self-lecture that goes with the arrival covers everything from the necessity for a heightened sense of positional awareness to the fact that the closer you get to the airport, the more traffic there is to spot.

If I have been to the airport before, I review the arrival procedures that I have encountered before. At Asheville, North Carolina, it's always like a dive-bomber attack on the airport because of surrounding high terrain. Likewise at Tucson. At home base, the opposite is often true: You often fly along at low altitude for miles. At Dallas, it's always a tour of east Texas. Wichita does it with efficiency. Those are just a few examples. As you fly around a lot, you learn that some facilities do a good job of mixing higher and lower performance traffic while others have a terrible time and manage only with excessive vectoring. If you know you are going to a place where they vector a lot, add time to the ETA and get there with extra fuel.

One final pilot item relates to fatigue. Because we are human, we fly in all states of alertness, from almost moribund to bright-eyed and bushy-tailed. Coming as it does at the end of a flight, the arrival finds us as far as possible from a good rest. Maybe we have to go to the bathroom, lunch was a Coke and Nabs, and breakfast . . . that was yesterday. To boot, we have been, for the duration of flight, operating at an altitude higher than that at which we probably live. In other words, it is not the most ideal of times for our hand-eye-mind coordination. But we have to make it the *best* of times for performance, because this is serious business. Accuracy comes from checks and double checks. "Track the 104 degree radial . . ." Set it and then ask yourself if you really set the 104 radial. "Cleared for the approach . . ." Does that mean it is safe to descend? Being a crew of one doesn't mean you can't have a challenge and response system for everything.

DRAW A PICTURE

When a pilot has a lot of trouble remembering to do all the things that must be done during descent, I like to have him draw that flight profile mentioned earlier. The line starts on the ground, goes up to cruise, and comes back down at the destination. Then I have the pilot write down where he will do all the various required things. This helps in two ways. One, it helps organize thoughts. Two, it graphically illustrates how the work load increases as the time remaining to fly decreases. There are a lot of notations in the descent phase of flight. Draw a profile, and you'll see what I mean.

APPROACH THOUGHTS

Things are set, the pilot is briefed, and the descent is planned. It's time to give some thought to the execution of the approach. If a descent is in rainy turbulence, it's harder to feel comfortable moving the old brain ahead 50 miles, but doing it will sure make an approach easier. Safer, too, because it is when you are 50 miles out that you set the stage.

Altitude is a key. There are several ways to know that we are flying at a safe altitude as we descend and maneuver to start an approach. Think ahead to the approach, and any special altitude considerations.

Published routes are the best havens. If you are flying on a published route at or above the minimum altitude, then the terrain and obstruction clearance is as assured in the arrival as it can be. When you are within 25 miles of the facility on which the minimum safe altitude is based, that altitude can be used. (Minimum safe altitude becomes minimum sector altitude if it is different in different sectors around the facility.) But beware the MSA. The facility on which it is based might, for example, be the outer compass locator. That has no distance information available — there's no way to know precisely when you are within 25 miles unless you have area navigation gear. And I'll always remember one case where there was a huge TV antenna located 21 miles southeast of an airport, 26 miles from the outer locator which was the basis of the MSA. The construction of this tower caused no immediate increase in the MSA, even though it was a rather clear hazard. (They have since raised the MSA, to reflect the presence of this tower.)

Another altitude is the minimum vectoring altitude. This altitude is available to the controller only, and it might be below a minimum en route or terminal route altitude, or below a published minimum safe altitude. But if the controller has a good target, he can clear the airplane to this altitude. I always like to have the lay of the land before descending to such an altitude and like to run a mental check as to whether I am where the controller thinks I am.

Knowing the lay of the lands helps. Some places we know from experience, from general knowledge of the area; others we might have to learn from a chart. Personally, I don't like to fly instruments without visual charts just in case some question arises about the terrain. I like to *know* where the tall towers and hills are located.

The classic accident that illustrates this is the TWA flight 514 crash west of Dulles, on the arrival portion of a flight. Not only did it result in a change in some air traffic control procedures (whenever they tell me to *maintain 2,000 until established* on the localizer, I think of 514 — that's where the terminology came from), but the accident also taught a lot of people a strong lesson about positional awareness and the lonely responsibility of the pilot-in-command to keep the airplane out of rock-lined clouds.

The flight was originally to have landed at Washington National. Because of strong crosswinds at that airport, flights were being held or diverted to Dulles International Airport, west of Washington. The crew had not operated there before.

During the descent from cruise, the crew discussed the instrument approach to Runway 12 at Dulles, the navigational aids, and the runway. Further discussions covered the various routings that might be expected. Then the controller cleared the crew to fly a heading of 090, to intercept the 300 degree radial of the Armel VOR, to cross a point 25 miles northwest of Armel, and maintain 8,000 feet. He told them that the 300 degree radial would be for a VOR approach to Runway 12 at Dulles. (Armel is located on the airport.)

The crew completed the preliminary landing checklist, and at 10:57 they again discussed the approach chart, including the Round Hill intersection which was an approach fix. The captain asked about weather between the flight and the airport, and at 11:01 the flight was cleared to 7,000 feet.

At 11:04 the approach controller said, "TWA 514, you're cleared for a VOR-DME approach to Runway 12." This clearance was acknowledged by the captain. At the time, the aircraft was 44 nautical miles from the airport. The captain said, "Eighteen hundred is the bottom," the first officer said, "Start down," and the flight engineer said, "We're out here quite a ways." In examining the approach plate effective at the time,

1,800 feet was the minimum altitude to be flown after passing Round Hill, which was located 17.6 nautical miles from the vortac. There were three published routes charted to Round Hill; all had minimum altitudes of over 3,000 feet. The flight was not on any of the routes. Additionally, obstructions of 1,930, 1,764, and 1,713 feet were charted to the west of Round Hill.

After some general discussion about turbulence and other factors, the captain said, "You know, according to this dumb sheet it says 3,400 to Round Hill — is our minimum altitude." The flight engineer then asked the captain where he saw that, and the captain replied, "Well, here. Round Hill is eleven and a half DME." (In error — it was 17.6, but this particular error was not critical.) The first officer said, "Well, but — " and the captain replied, "When he clears you, that means you can go to your — " An unidentified voice said "initial approach," and another unidentified voice said "Yeah!" Then the captain said, "Initial approach altitude." The flight engineer then said, "We're out a — 28 for 18."

At 11:08, the flight engineer said, "Dark in here," and the first officer stated, "And bumpy too." Then the sound of an altitude alert horn was recorded. The captain said, "I had ground contact a minute ago," and the first officer replied, "Yeah, I did too." Then the first officer said, " — power on this — " and the captain replied, "Yeah, you got a high sink rate."

There were a few comments about turbulence, then at 11:08:57, the altitude alert sounded. Then the first officer said, "Boy, it was — wanted to go right down through there, man," to which an unidentified voice replied, "Yeah!" Then the first officer said, "Must have had a [expletive] of a downdraft."

At 11:09:14, the radio altimeter warning horn sounded and

stopped. The first officer (who was flying) said, "Boy!" At 11:09:20 the captain said, "Get some power on." The radio altimeter warning horn sounded again and stopped. The sound of impact was recorded at 11:09:22. The airplane hit the side of a mountain at an elevation of 1,675 feet.

The NTSB found the probable cause to be the crew's decision to descend to 1,800 feet before the aircraft reached an approach segment where that minimum altitude applied.

In a much less publicized accident not too many miles away, an airline pilot flying a general aviation airplane hit a ridge quite a distance from the airport while on the final approach course to a different airport. Same story, and you can't help but wonder if the spoken words of TWA's crew were the thoughts than ran through the other pilot's mind as he rationalized the altitude to which he was descending.

When a professionally flown airplane has such trouble on an arrival, it is a reminder to all of us that just the act of planning or discussing things doesn't satisfy the requirement to double-check everything. If there is any question, it should be resolved. Nobody will ever know why the TWA crew did not scrutinize the approach plate more carefully when the captain questioned the descent to 1,800 feet. If they had seen the terrain depicted on the chart at levels near or above that value, they surely would not have descended. If they had had a knowledge of the terrain, if they had flown over the beautiful Virginia countryside on a clear day and stored the scene for future reference, they would have known that there is a ridge out there. "If" doesn't resurrect people and airplanes, but it can sure help to keep others out of the weeds and trees. In this case, the crew was cleared for an approach while a considerable distance from the airport, which is unusual, and they started the approach while still in the arrival phase of the flight.

HOLDING

Holding is part of an arrival, and is actually something to be avoided if at all possible. Holding is an unnatural act, and very much of it is a sign that things are about to get fouled up. Whether the hold is for traffic or for improvement in the weather, it can only cost fuel and nibble away at the margins that have been built into the flight. Taken to an extreme, holding can lead to greatly increased pressure on the pilot to complete the approach — perhaps because there is no other alternative.

An accident involving a commuter airline Twin Otter offers an example of problems created by arrival holding.

The aircraft was scheduled for a fifty-minute flight into New York's Kennedy Airport. About twelve minutes after takeoff the crew received a holding clearance. The controller said there was a possibility of no holding. Then the controller told the crew that if holding were to be required, they would have an expected approach clearance time of 17:06, thirty minutes in the future. The aircraft entered the hold and established contact with Kennedy approach control at 16:48.

At about this time, the weather deteriorated, and aircraft began missing the approach to Runway 13L. So they switched runways, to 22R, which had better runway visual range. Because of the switch, the Twin Otter was cleared to another holding pattern and was given a new time to expect an approach — 17:46. At 17:27 the approach controller attempted to radar identify the aircraft, but the aircraft transponder was apparently not working. The controller then told the crew that it would be an approach without radar service, which meant greater separation between where he thought the Twin Otter was and other aircraft. At 17:39 the controller told the crew that "in about fifteen minutes it will be okay."

Then, at 17:42, one of the crewmembers said, "Request New Haven if we can get there, expedite." The clearance came promptly, and they headed off toward New Haven. The New Haven weather was 200 scattered, 500 broken, 700 overcast, with 2 miles visibility in rain and fog.

As the flight neared New Haven it was, you guessed it, cleared to hold and was given an expected approach clearance time of 18:15. Then at 18:12, the flight was told that, based on a report from a pilot who just landed, the airport was below minimums. (The question of how that pilot landed remains unanswered.) The crew asked for the Bridgeport weather and was told that it was worse than New Haven, so they decided to take a shot at the latter despite the pilot report.

Shortly before 18:17, the crew reported a missed approach. Back to approach control, and to new holding instructions. The crew asked for the Groton, Connecticut, weather. It was 300 overcast with 1½ miles visibility. Finally, at 18:20, the crew shed light on the seriousness of their problem: " . . .We got minimum fuel now, we gotta get to Groton." The crew then elected to head toward Groton before receiving an IFR clearance to do so. (They said they were VFR on top at the time.) Next they requested an approach direct into Groton. The controller gave them a frequency change, but they were unable to contact that facility.

At 18:22 the crew called a flight service station and explained the communication and fuel problem. The FSS procured the desired clearance and gave it to the flight.

At 18:24, the captain of the flight called his company radio and said, "We're coming home with this pile of junk. We ain't got any equipment working." The captain then had a discussion with the chief pilot and was advised that the Groton weather had lowered to 200 feet. The captain reported that he was inbound, flying at about 100 feet over the water. Asked how much fuel he had remaining, the captain answered "none." Then he radioed that he had lost one engine and was trying to make shore. He then reported that he had lost the other engine, couldn't make shore, and was going into the water. This was at about 18:37, or one hour and thirty-six minutes after the first expected approach clearance time issued at Kennedy.

As you might expect, the probable cause issued by the NTSB was "fuel exhaustion resulting from inadequate flight preparation and erroneous in-flight decisions by the pilot-in-command."

It is in the arrival part of a flight that we get the bad news about holding, and it should always be dealt with decisively. I remember hearing a controller save a pilot's bacon one night and always wondered what the pilot was thinking about as he entered holding in a business jet because the weather had gone below minimums at his destination.

It was dark, and the weather had gone to zero-zero. When told, the pilot had matter-of-factly replied to the controller that he would hold for improvement. A minute or two later the controller, who knew the local weather and knew that it would probably stay zero-zero all night, asked the pilot how much fuel he had on board. The reply was "forty-five minutes." The controller then took matters into his own hands. As best I recall, he said, "Well, you can hold if you want, but if I were you I'd hustle to Baltimore. It's above minimums but won't stay there long, and we'll probably be socked in all night." The pilot awakened, was cleared to Baltimore, and lived happily ever after. At least he lived through that night.

EXCEPTIONS

Sure, there are times when it's okay to hold for improvement in the weather and there are times when we have no choice but to hold because the controller says we have to hold. The key is always related to fuel. If the tanks are awash with the stuff, then holding will work for a while. But it can't be allowed to be a mindless exercise. When first cleared to hold, we should strongly question the wisdom of continuing on with the original path.

One evening I was inbound to my home base, Mercer County Airport in New Jersey, with deteriorating weather being reported despite an optimistic forecast. It was at the end of a long week of travel, and I was ready to get home to the fireside. The leg being completed was over 700 nautical, so there wasn't a lot of fuel in the tanks. The legal amount of fuel was there, but with weather worse than forecast I didn't feel at all adventuresome. And when they called with word that the proposed destination airport now had weather that was reported below minimums, it took no time at all to make a decision. I would not try the approach, and I would not hold for improvement. To do either would be foolhardy. Instead, the landing was at Philadelphia International, which had landing minimums. Even with a lot of fuel, holding for improvement or trying the approach might not have been the smartest things in the world to do. I have enough knowledge of the area weather to know that on a night with deteriorating weather the best place to land is always the closest place with minimums. Procrastination can only serve to increase the distance between your airplane and places with above-minimum weather.

Holding for traffic during an arrival is also something to question. How involved to become in this depends largely on the situation of the moment. If traffic is backing up because of weather — thunderstorms in the area or weather fluctuating above and below minimums — chances are it will take the controllers longer to work off the backlog of traffic than they anticipate. I wouldn't bet a nickel on any expected approach clearance time. And remember, even being legal and having enough fuel to go to the alternate and fly for forty-five minutes doesn't mean a lot. If for example, you hold for thirty minutes, shoot the approach, miss it, and have to go to the alternate, that would mean a landing with fifteen minutes worth of fuel — hardly a desirable event.

If flying with anything close to the legal minimum on fuel, I think that I'd choose to divert at mention of the word *hold*. Flexibility is one of the strongest assets of the general aviation airplane, and this is a place where a singlemindedness of purpose — getting to the filed destination in this case — can lead to critical situations.

I'm not going to offer any words of refreshment on entering a holding pattern. It almost never happens in the real world and the few that I enter in the course of flying IFR all over the country are easily managed with logic.

It is true that there is a lot to do and to think about during an arrival. It is also true that doing it and thinking about it in this phase of flight makes easier and safer the next chore — that of the approach and the landing.

CHAPTER 6

Approaches

FIRST PART   Patrick E. Bradley

Five hundred feet to go; the airplane is established on glideslope and on localizer. . . . 300 feet to go; speed looks good, heading and descent rate are on target. . . . 100 feet to go; the airplane is stabilized; 50 feet . . . 10 . . . and decision height. You look up and there they are, a welcome sight to any pilot — the flashing strobes of the approach lights, inviting you to complete the approach and the flight. Taxiing back to the ramp, raindrops peppering the fuselage, it's hard not to feel a certain sense of satisfaction in successfully completing the flight. An instrument approach to minimums is a demanding task, one that requires a pilot to muster all his skill and judgment. More than at any other time during the flight, momentary lapses can become serious errors, and I suppose this is why most pilots, training for their instrument rating, spend a vast majority of their total lesson time flying approaches.

One reason that instrument approaches are often difficult to learn, at first, anyway, is because they tend to reveal latent defects in flying technique. After flying more bad approaches than I'd like to remember, I've learned that the key to a good instrument approach is to first master basic attitude instrument flying skills. With a sound grounding in the basics, an approach

142

becomes just a challenging exercise. Without the basic foundation, a good approach can be nearly impossible.

When I began my instrument training, I flew approaches backward. Once I had intercepted the localizer and the glideslope on an ILS, I focused all of my concentration on the glideslope and localizer needles, bothering with the DG, artificial horizon, or vertical speed indicator only to see how much correction I had in at a particular moment or how much correction I was applying to catch the wild swings of the needles. I never had much of an instrument scan for approaches, I never needed one — the little needles told me everything I wanted to know. Needless to say, that technique was far from successful. My instructors were constantly admonishing me for "chasing the needles," but none of them ever told me *why* I was chasing the needles. I never learned until much later that my primary problem was flying the approach solely in relation to the localizer and glideslope deviation indicators.

Once I began to fly instrument approaches with reference to the artificial horizon, the directional gyro, and the vertical speed indicator, I began to make some real progress. By limiting my direction changes to five degrees maximum and my pitch changes to one bar width, I was able to take a calculated and consistent approach to tracking the localizer and glideslope. If I needed to correct for changing winds, I knew how much I had corrected and how much correction I would have to take out to maintain the proper heading and descent rate. Finally, I was able to control the needles instead of having them control me. It was a big step. My approaches are still far from perfect, but I always do best when I concentrate first on flying a heading and maintaining a proper descent rate, using the glideslope and localizer needles for guidance.

Beyond the basics of good flying technique, there are operating procedures that simplify approaches. Assuming that all of the radios are set and the checklists substantially complete at the marker, the approach is a fairly straightforward proposition, whether it is a VOR, a NDB, or an ILS. But certain memory items are crucial. The inbound heading and the MDA or DH are primary. Just looking down at an approach plate in the midst of an ILS is enough to throw into disarray the whole process of maintaining heading and descent. One of my instructors always took my approach plate when I hit the final approach fix. When I knew that the plate wasn't going to be there, I almost never forgot the numbers. When I had the information in front of me, though, I would constantly look back down at the information — just to make sure. Another helpful practice is to use the heading bug on the directional gyro to pinpoint all headings. I've gotten into the habit of using the bug for all headings, but it's especially useful during the approach. By doing this, it's possible to see at a glance whether I've drifted from the heading I established for the wind-drift correction. Basically, using the heading bug simplifies the task of holding a heading. Also, I set all assigned altitudes in the moving card of the ADF. It's just a reminder, but it's saved me more than once. Just the act of setting the altitude to the corresponding number fixes it in my mind; it's a worthwhile habit to pick up.

After hitting the final approach fix with all of the memory items firmly in mind, the next step is to get the approach stabilized. With any luck, I am generally pretty well established on the localizer by the time I reach the outer marker on an ILS. If that's the case, then the only task is to establish the airplane in about a 500-foot-per-minute descent by lowering the gear, reducing the power to the proper setting (pilots usually figure this out beforehand with some experimenting), and lowering one notch of flaps. Retrim the airplane (some, I've found, don't even need to be retrimmed) and presto, the airplane is established in a 500-foot-per-minute descent. From that point on adjustments become the primary concern.

Unfortunately, it's easier to describe the ideal approach than to fly it. I have firsthand experience, as do all pilots, with the difficulty of establishing an airplane on approach and keeping it there. Winds change, updrafts and downdrafts wreak havoc on the descent, and sometimes the trusty power setting just doesn't make it. Adjustments become not just a matter of monitoring the airplane on its approach to the threshold, but guiding the airplane along the proper approach path without bullying it all over the sky. Bullying the airplane is the result of heavy-handed adjustments. The same instructor who took my approach plates from me often threatened to administer the book to the side of my head if I even considered correcting more than 5 degrees in direction or an artificial horizon bar-width in pitch. His advice worked. But there is always the temptation to return "quickly" to the localizer or onto the glideslope. Yielding to the temptation usually means that I will cross the glideslope or the localizer often, never making more than a passing acquaintance with the proper path.

After understanding the concepts of flying a good approach, I think the most important step to flying a good approach is practice. I find that when I've practiced approaches, I can fly them well. I can judge the proper degree of correction more accurately when I am sharp, and the procedural steps become second nature. With enough practice, one step follows another, and the airplane flies the glideslope and localizer with a minimum of correction and hunting for the proper heading and pitch attitude. Unfortunately, I find that when I have not practiced enough, I revert to the same bad habits I had overcome earlier. Flying an instrument approach is not the same as riding a bicycle. It's easy to lose the touch, no matter how many hours or approaches you have under your belt.

Practicing approaches requires another pilot, and even if he isn't an instructor, he may be able to focus on your problems if he is also instrument rated. Of course, his main job is to keep a lookout for other airplanes, but he may also be able to offer an objective evaluation of the approach above and beyond your own critique. Sometimes I analyze my own performance as I fly the approach. I "talk" myself down. From the final approach fix in, I give a running commentary of what the airplane's situation is at a given moment and what I will do to remedy the problem. Sometimes I see a problem — say the airplane is drifting off the localizer. If I verbally respond to the error as I see it happening, I avoid fixating on the instrument without making a correction. Also, after I recite the problem and the remedy, and after I act on it, I know that I should be moving on to the rest of the panel. I know it sounds goofy, but the process reinforces the pattern of evaluating an instrument, correcting, if necessary, and moving on to the next instrument. Without exception, I always fly a better approach when I talk myself down. I often wonder why this isn't an established practice, at least in initial instrument training.

Weather on the approach is a significant consideration; often, it will determine how you fly the approach, or whether you fly it at all. For example, frontal activity in the vicinity of the airport is likely to affect winds as the airplane descends on the approach. Whereas the approach began with a tailwind, it may change to a headwind, affecting airspeed and descent rate. Increased airspeed, increased rate of descent, and wind shear will be considerations throughout the approach. Thunderstorms in or about the vicinity of the approach or the missed approach path are a good reason not to shoot the approach, as far as I'm concerned. Approaches can only go in one direction if they are to end successfully, and that means that diversions, except to another airport, are pretty much out of the question.

Icing on the approach is another frequent concern. Even light accumulations affect the handling characteristics of most airplanes, but it's difficult to be certain how. Pilots know that the approach may require more power, and that flaps should not be used. Beyond that, few pilots have ever given me any conclusive advice on how to fly an approach in icing conditions. One pilot once told me that above all, I should not rely on past experience with the airplane's normal handling characteristics. Chances are good that they may be entirely changed. Also, I would be very reticent to continue an approach into icing conditions when there is a chance that I might have to miss the approach. The airplane just might not be able to climb out with a load of ice.

Night approaches can also present unique problems. I think they are tougher for several reasons. If I'm flying a night approach it's generally after a full day, and fatigue becomes a primary concern. I find that I've got to double- and triple-check actions that I generally get right the first time. Frequency numbers somehow get transposed at night, approach courses get reversed, and sometimes just holding heading and altitude can become a challenge. Without constantly pushing my scan, the temptation to fixate on a single instrument becomes almost uncontrollable. Then, there is the added inconvenience of reading charts in the shadow of the airplane's red overhead light.

Another troublesome aspect of night IFR is the landing itself. Breaking out of a low overcast and landing the airplane in daylight can be tricky, but doing the same at night is twice as difficult. Without the day's horizon I generally flare about ten feet higher than I ought to. As a result, I make firm landings, but that at least wakes the sleeping backseat passengers. My night instrument approaches are often characterized by the ongoing process of making and, I hope, correcting minor mistakes. I have never flown night IFR alone, and I think I would go to great lengths to avoid that challenge.

Different approaches impose different demands on the pilot, but they are all similar in that with good flying skills —holding heading, altitude, and speed — they are no more than variations on the theme that characterizes all instrument flying. My instrument instructors never concentrated on the basics of attitude flying in any express manner. In fact, it wasn't until after I had gotten my rating, when I took an instrument refresher course at FlightSafety International, that I really began to understand its importance to instrument flying. Attitude instrument flying gives the pilot a starting point on which to build solid instrument skills. Once he's mastered them, an ILS, VOR, or an NDB approach becomes a matter of learning the procedures and practicing them. It was only after returning to the basics that I ever felt I had control of my airplane on an approach; I only wish that I had learned it earlier.

SECOND PART   Richard L. Collins

Whether it's your first approach since getting rated or your hundredth, the time of starting the final descent is a time to remind yourself that this part of the flight demands absolute precision. It also is almost always a time of change. If the weather system is creating low conditions, this fact might mean that there will be some wind shear within a couple of thousand feet of the ground, which could result in some turbulence and affect the airspeed and sink rate.

If flying an airplane other than the simplest, it means we'll be operating a "different" airplane than the one we have been flying. The cruise portion was at a higher airspeed than the approach. On final, the gear and at least a portion of the flaps are deployed, resulting in a big change in the flight characteristics of some airplanes and a small change in some others — but change nonetheless. Add these changes to differing weather conditions at low altitude and to the quickening pace of events, and the demands as an arrival evolves into an approach and the approach evolves into a landing are clear.

The quickening pace: Things might become a bit much for one pilot on one approach and for another on a different approach. I'll offer a couple of examples.

It was a dark and stormy night, and we were coming home after a long day of flying. My son Richard was in the left seat. He didn't have his instrument rating at that time, but he had been flying instruments since he was big enough to reach the controls. (Before that, he had to be satisfied with a desktop simulator.) The tower was closed at home, but the last reported weather was close to minimums and other stations in the area were reporting similar conditions. A front was in the process of passing, it was raining hard, and the wind was shifting to the northwest and freshening. As we discussed the approach during the arrival phase, I told Richard that I wanted a good performance. Needles crossed all the way in — no missed approach because of anything other than weather. It was late, and doing it exactly right the first time was essential as far as I was concerned.

The wind was out of the southwest as we descended; from surface reports we knew that it would shift to northwesterly as we descended on final. The approach was to Runway 6. We reviewed the principle that a decreasing tailwind makes the airplane trend high on the glideslope, and that the airspeed will stay on the high side as the airplane adjusts to the new wind direction and velocity. Also, there would be some turbulence and an increasing crosswind from the left.

He said he was cocked and ready so I let him carry on. He did an acceptable job of keeping the ILS needles in the center as we descended. The increasing drift was impressive to him, and as we started breaking out at 400 or 500 feet, the visual picture was one of a strong crosswind. He looked up, the runway was in clear sight, but the airplane had about 15 degrees of drift. Rain was streaking the windshield, and turbulence was buffeting the old airplane. He said, "Okay, Dad, now you can have the landing." The generous fellow had seen a bit much. The main thing is for him to remember that experience, and to avoid flying into such a situation until he is ready to handle the crosswind landing at the end of the approach.

Another pilot, seasoned but unfamiliar with my P210, came on a different situation. He was working an approach to minimums, gear and approach flaps down, and was doing a fine job of flying the airplane. We'd have calm conditions for landing, and he was a little fast but the runway was quite long. He was trending a little high on the glideslope, and when we were at about 600 feet above the ground, still in the clouds, the tower announced that they were launching an airline twinjet on our landing runway — "caution wake turbulence."

That changed the equation. I had decided that a decreasing tailwind was making us tend toward being high on the glideslope, we were fast, and only approach flaps were extended. All that had been okay, but with a jet lifting off, a landing far down the runway suddenly seemed like a bad deal. We'd want to be down well before reaching his liftoff point. I quickly announced that full flaps were going out, and moved the selector. The 210 pitches up strongly with flaps extension. This pilot wasn't ready for it, and I had to help him just a bit. The pace had quickened at a bad time and, to him, in an unexpected manner.

STABILIZED VALUE

In the first approach, Richard's, the airplane was stabilized in the approach configuration and was flown that way on final. In the second, the configuration was changed on final. In the first approach, we had no control over the changes that occurred. In the second, we did. And in the second case, it would have been a lot easier on the pilot flying if we hadn't changed configuration. He would have shot a perfect approach. The stabilized approach does indeed have a lot going for it, and we should strive to stabilize all of them at the final approach fix. But we have to be able to handle different situations.

I'll always remember watching an excellent stick and rudder pilot fall apart while trying to accommodate a controller's requests. The controller had worked himself into a box, and the only way his traffic situation was going to work was for us to intercept about a mile outside the marker and to maintain as much speed as possible. He told us that this would be the case, and we continued.

I have a procedure to use when the request is to maintain speed, and I quickly explained it to the pilot flying. My explanation must have gone over his head for halfway from the marker to the airport, during the course of rather aerobatic lunges at the ILS, the pilot announced that he "couldn't do this." It was his feeling that the whole thing was unreasonable, and that we should have refused the request.

Perhaps we should have. Certainly you don't ever want to attempt something that you don't feel comfortable doing. But we live in this world, not the perfect world, and the better we are able to adapt to the foibles of the air traffic control system in which we fly, the better that system will work for us. Approaches are where we are most often called on to change the routine. The change is usually related either to speed or to the intercept of the final approach course. Success is found in having a plan for every eventuality.

When we sift through the accident reports, though, few of the probable causes are directly related to a pilot's having trouble adapting to the system. It's not an immediate safety of flight item, although it probably is an indirect cause of some problems because of distractions. Most of the IFR accidents do occur on approaches, but most of the approach accidents relate to the pilot leaving a safe altitude without the runway in sight. When we have trouble with the system, we can usually patch that up. Miss the approach, tell the controller that we can't maintain that much speed, or request a turn onto final at least five miles outside the marker. With such requests, we might well become a pain in the tail to the person on the ground. But when we miss the clear signals that an approach isn't going to work and we keep trying to make it work, then the imminent pain will be ours.

DESCENT BELOW MDA

An airline accident at New Haven, Connecticut, reveals how the approach can turn lethal when the emphasis isn't on the published altitudes.

The sky was partially obscured in fog at the airport. The

visibility was 1³/₄ miles. According to the NTSB report, the flight had previously missed approaches at another airport. On each of the missed approaches the crew had descended well below minimums, according to the flight recorder. On one, the airplane was but 125 feet above the ground when the missed approach was commenced. (The minimum descent altitude for the approach was 610 feet.) The aircraft finally landed at that airport, passengers were unloaded and loaded, and the flight took off and headed for New Haven. The transcript of the cockpit voice recorder tells the tale:

APPROACH CONTROL: Four eighty-five, roger, Bridgeport, make it New Haven weather, sky partially obscured, 1³/₄ in fog, wind 180 degrees at 5, altimeter 29.97.

FIRST OFFICER: Okay sir, and we'll take a turn-on right at the point if that'll be all right.

APPROACH CONTROL: Okay, you want to turn right into the airport?

FIRST OFFICER: Yeah, that'll be okay at, ah, Pond Point be fine.

APPROACH CONTROL: Okay, I didn't get the latter part of that, turn right heading 360.

FIRST OFFICER: Three six zero, Allegheny 485 turning.

APPROACH CONTROL: Okay and you can intercept the final approach course on that heading. You'll be right at Pond Point, you're cleared for a VOR approach.

FIRST OFFICER: Okay, thank you, sir, cleared for the approach, Allegheny 485.

APPROACH CONTROL: Roger, you're welcome.

CAPTAIN: What was that wind, you, you remember offhand?

FIRST OFFICER: A hundred and eighty at 5.

APPROACH CONTROL: Okay and Allegheny 485 contact New Haven Tower 124.8 now, you're Pond Point inbound.

FIRST OFFICER: . . . 'kay, twenty-four eight, thank you a lot.

APPROACH CONTROL: You're welcome, sir.

CAPTAIN: Before you talk to him, will you give me 15 and gear down please.

FIRST OFFICER: There ya go.

FIRST OFFICER: New Haven Tower, Allegheny 485 is, ah, passin' the point comin' inbound.

(The Pond Point intersection was 5.7 miles south of the airport; the minimum descent altitude was 380 feet mean sea level.)

TOWER: Roger 485, runway your choice sir, the wind 190 degrees at 5, altimeter 29.96, Runway 2 or 20.

CAPTAIN: Well, tell him, ah, well that's all right, we'll take 2.

FIRST OFFICER: Okay, the way it looks we'll take 2, be all right?

TOWER: Roger, cleared to land Runway 2.

CAPTAIN: Out of a thousand.

FIRST OFFICER: Twenty-four?

(The final checklist items were handled next on a challenge and response basis.)

CAPTAIN: Give me 40.

CAPTAIN: I'm tellin' you.

FIRST OFFICER: Out of 500 . . . looks like about 100 feet atop.

CAPTAIN: They sure do.

UNKNOWN: Not very good, is it?

FIRST OFFICER: Top minimums. . . . I don't have it.

FIRST OFFICER: Decision height . . . you got 105 sinking 5.

CAPTAIN: All right . . . keep a real sharp eye out here.

FIRST OFFICER: Okay. . . . Oh, this [expletive] is low. . . . You can't see down through this stuff.

CAPTAIN: I can see the water. . . . I got straight down.

FIRST OFFICER: Ah, yeah, I can see the water. We're right over the water. . . . Man we ain't 20 feet off the water. . . . hold it.

The first officer's "hold it" came about one second before the aircraft flew into three houses on a beach about a mile south of the airport.

It is tempting to dismiss that chain of events with the thought that you would never descend below minimums. If your self-control is that good, fine. But I think that almost all of us occasionally fly with that little pitchfork-carrying fellow on our shoulder. At least, we are occasionally tempted to sneak down just a little lower for a better look. What the transcript of this accident teaches us is the folly of descending below the minimum safe altitude in the belief that we'll see something. This might be true in a rain situation, but it is far from true in a fog situation. Ground witnesses in the area where this airliner crashed reported that the visibility was but 50 feet over the water and from 150 to 200 feet over the beach area. It's a simple fact that in fog, the lower you go the less you see. And it is true that as you progress in an approach, the pace quickens and the margins of safety decrease. Any departure from the published procedure or altitude can only compound this. As long as the procedure is followed, at least the pace quickens in a more or less predictable manner. If you deviate from the procedure, it's not possible to anticipate what might come next. We saw that on an arrival with the example of TWA 514: descending below the minimum descent altitude led to a house on the beach. If the procedure is flown, the worst it can lead us to is a missed approach.

CONSIDER ALL FACTORS

The key to safety and a smoothly flown approach is an awareness of the factors that will affect the airplane on the approach, plus a continuous awareness of position, of what is happening now, and of what must happen next for things to go according to plan. If you have a plan for everything, it makes life a lot easier. And the plan has to be flexible. Some approaches begin in one manner, some in another. The arrival should have gotten the aircraft to the final approach fix on-ready; you too should be ready. If you *know* the power setting and configuration and pitch attitude to track the glideslope, and know the bearing of the final approach course plus an estimate of drift, the approach can begin with everything as close to on track as possible. If the descent and the heading are nearly right to begin, the only adjustments required will be to compensate for wind. The plan, again, is to get the airplane started in the correct direction, after which you fine tune. Wind shear has the biggest effect on our fine tuning, and it should be considered on every approach.

If there is to be wind-shear effect on the approach, we can anticipate this with but basic information.

To begin, be aware that the wind-shear detection equipment the FAA has installed is really wind *shift* detection equipment. It compares wind sensors around the periphery of the airport with the primary equipment on the tower; if there is a difference, an alert is given the air traffic control personnel. It does not provide any warning when there is a substantial difference in the wind within 2,000 feet of the surface and the wind at the surface. It is that shear in the lower altitudes that can have an effect on descent rates and airspeed on final as well as on the heading required to fly the approach.

When flying the arrival, we should have an idea of the wind in the lower altitudes based on groundspeed and drift. The surface wind is available from the automatic terminal information service or from some other source. If there is a substantial difference in wind aloft and wind at the surface, then the pilot needs to be alert for wind shear. The most common condition encountered in the U.S. is a strong southwesterly wind aloft and a light or an easterly wind at the surface. If you make the effort to develop a mental picture of the air through which you are about to descend, surprises are more likely to be avoided.

It's all been said before, but a refresher on the effects of wind shear on approach is always time well spent.

From the beginning we are taught that, once free of the ground, the airplane operates in the air mass and that the wind affects only the track over the ground. Where this isn't true is when the wind aloft is variable — either over a vertical or horizontal distance. When the wind changes over a relatively short distance, one result is turbulence, caused when air moving at different velocities rubs together. The other effect that it has on the airplane is, oddly enough, best related to groundspeed.

Consider an airplane descending on an ILS approach to Runway 6, with calm surface winds and a 50-knot southwesterly flow at 2,000 feet — not an unheard-of situation. If the true airspeed of the airplane is 120 knots at 2,000 feet, the groundspeed of the airplane will be 170 knots. In the four minutes that it takes to descend that last 2,000 feet, the airplane will make a transition to a condition with no wind, and if the airspeed remains constant at 120 knots, the groundspeed will have to decrease by 50 knots. That does not happen without some effect on the airplane. A 50-knot deceleration in groundspeed is required in two minutes, and you know what has to happen: Reduce power and adjust the pitch attitude. Or, put the other way, a decreasing tailwind on an approach will make the airspeed trend high and the airplane trend high on the glideslope. A decreasing headwind would do the opposite; the airplane would have to accelerate on the approach to maintain an indicated airspeed. In the example, the groundspeed at 1,000 feet with a 50-knot headwind would be 70; it would have to increase to 120 before the airplane reached the point where there was no headwind. In heavy airplanes, where acceleration is critical in the approach configuration, a rule-of-thumb is never to fly with the groundspeed less than a safe indicated airspeed for the approach.

Two other factors bear on this. One — a light airplane handles these accelerations and decelerations much better than a heavy airplane in the approach configuration. The heavy airplane is in a high-drag condition and is already using a substantial amount of power. In most light airplanes, the drag isn't so high. Not much power is used on a normal approach, so there is plenty in reserve; and the runway ahead is usually long, so we can pack on some extra airspeed if the condition is one that might cause an airspeed decay. The other factor is that the primary change in wind might occur over a relatively small vertical distance, in some conditions. I've heard pilots report a 10- or 20-knot effect on airspeed on short final, meaning that there is a strong and primary shear at 100 and 200 feet that has to be handled instantly. Again, light-airplane pilots have an easier time with wind shear than heavy-airplane pilots. And all pilots can anticipate shear by comparing the low-level wind aloft and the surface wind.

TRY AGAIN

Wind can do other things to us on approach. An accident involving a Citation business jet shows how wind (there was apparently little or no shear this day) created a situation where an approach turned out to be a mission impossible.

The weather report furnished to the pilot before he started the approach was 700 overcast, visibility 1 mile, light snow and fog. The surface wind was reported to be from 70 degrees at 13 knots; the braking action was reported as poor. The ILS at the destination airport serves Runway 23; the minimum descent height for a circling approach was 543 feet; the visibility minimum for the approach was 1 mile.

While almost all pilots will agree that a circling approach in minimum conditions is a bad deal — it's nothing but scud-running around the airport when the reason we fly IFR is to avoid scud-running — almost all of us would rationalize a try at the approach as it existed that day. The conditions were at or above minimums, and it was daylight — there's a big difference in trying something like that in the daylight and in the dark.

The crew flew the approach and apparently saw enough of the airport to set up on a downwind leg for Runway 5. The airplane was sighted flying parallel to that runway; then it was seen to enter cloud. As is the proper procedure, the crew executed a missed approach. They then requested another ILS.

On the second approach the crew elected to land straight in on Runway 23. The runway was 4,742 feet long, and, according to the National Transportation Safety Board report, the last wind given the crew was 070 at 13.

Flight service station personnel on the ground stated that the second ILS approach looked normal and the aircraft appeared to touch down about 500 to 1,000 feet from the end of the runway. Another witness said that the airplane appeared "a little high and coming in faster than normal" during the approach. Two witnesses said the airplane appeared to touch down about halfway down the runway. One witness reported observing the flaps retracting before the aircraft reached the last taxiway, and all witnesses heard an increase in engine power and saw the aircraft rotate for takeoff before it ran off the end of the runway with disastrous results.

The runway was icy, and the NTSB calculated that, based on the icy-runway correction factor of 100 percent in the aircraft operating manual, the aircraft would have required more than 5,250 feet to stop on the downwind landing. That exceeds the runway length.

LACK OF INFORMATION

In general aviation IFR flying, we often fly to airports where there is no weather reporting, no surface wind reports, and no runway condition reports. So we might find ourselves faced with a similar approach with even less information than was available to this crew. That's no reason for the event to have other than a safe conclusion.

First, a downwind landing should be contemplated only when the available runway length is grossly in excess of that required by operating handbook calculations and when the downwind component is below the maximum listed. Second, if the surface is tricky — wet or especially ice — the retreat to another airport should be started sooner rather than later. Some pilots will touch down to test braking action; I might try this but only if I'm landing into the wind with a lot of runway ahead. I'd pass on doing it with a downwind component.

Finally, weather reporting: This is often mentioned, and I often wonder what it has to do with it. An instrument approach is a cut-and-dried event. You descend to the MDA or DH, and if certain visual flight qualifications are met at that point, you continue and land. If the visual requirements are not met, you go away. There is nothing dangerous about descending to the published minimum and then flying the published missed approach. So I don't understand the FAA's ban, for example, on air taxi approaches to airports where there is no weather reporting. And I don't understand reference in any accident report to a lack of weather reporting. Thinking back to that last accident related, the weather report did the crew precious little good. The report suggested suitable conditions for a circling approach; that was the old come-on. The pilots tried, the report was wrong, so they tried something else. The outcome couldn't have been worse with no reporting.

NO APPROACH

In general aviation operations we often have to up the ante on that lack of information before an approach and landing. There are times and places where we turn an IFR flight into a VFR flight for an arrival at an airport without a published approach. This can be a very delicate business and is a factor in a number of IFR accidents.

A good example is found in the NTSB report on an accident at Sun Valley, Idaho, involving a Challenger business jet.

Sun Valley is a beautiful spot, but the airport is not one of the most unobstructed in the world. It is in a valley, and the mountains around are of generous size. The accepted method of operation is to land to the northwest and take off to the southeast. The terrain rises to the northwest of the airport; to the southeast the valley is wider and the terrain becomes reasonably flat in a relatively short distance. The recommended arrival procedure is to approach from the southeast VFR, flying the east side of the valley at 7,000 feet mean sea level until the runway is in sight. (Field elevation is 5,315 feet.) There's a town on final approach; the procedure recommends that pilots fly over the east side of this town to intercept the final approach path to the airport, according to the NTSB. A railroad track and highway run from this town to the airport; the runway is parallel to the road and to the tracks.

The pilots of the Challenger had been to Sun Valley before and were familiar with the area. This day they canceled their IFR flight plan at 17,000 feet. The last recorded radar position was right over the airport at 13,500 feet. At about this time the crew called Unicom with word that they were over the field at 7,000 feet (the altitude reference was apparently to above ground level) and relayed routine requests for food and fuel service.

Other pilots arriving and departing estimated the ceiling at the airport to be from 800 to 1,500 feet and visibility to be from 3 to 10 miles. Ceilings and visibility reportedly lowered north of the airport. Pilots' reports indicated that the overcast was thinner to the south and that there were breaks.

No definite flight path could be established, but based on two witness sightings of the aircraft it was below clouds west of and abeam of the airport. The airplane was within "300 to 500 yards" of the mountains to the west of the airport, and it then turned right and "disappeared" into "low hanging clouds" over the northwest side of the airport. It was next seen east of the town of Hailey (which is just northwest of the airport). It was still below the clouds. The airplane subsequently crashed into a mountain 2.2 nautical miles north of the airport. Given the described cloud bases and the observations, the NTSB concluded that the airplane was flying on a northwesterly heading as it approached the town of Hailey and aligned along the western side of the valley. Then it began a northerly turn, passed to the north of the Sun Valley airport over the southeastern tip of Hailey, then east of Hailey and toward the crash site. A right turn apparently continued until the aircraft reached the impact heading, approximately 15 degrees magnetic. The calculated elevation at the impact site was 6,520 feet.

The NTSB found the probable cause to be the crew's failure to adhere to the recommended visual approach procedures, but it could not conclusively establish why this was the case. The Board did find that obscuration of terrain features and landmarks by snow made navigation by visual references, and terrain avoidance, difficult.

The Board did say that the crew might not have definitely established its position visually as it flew toward the airport and that, as a consequence, it might have misidentified the town of Hailey as the town 4.4 miles to the southeast. This would have led them to believe they had farther to fly to get to the airport. The Board also theorized that the crew might have been using an onboard long-range navigation system to help in locating the airport, and that any error there could have reinforced the crew's belief that Hailey was the other town.

LESSON

The lesson here is not new but is always worth additional thought. Certainly all pilots want to deliver the goods, and if other aircraft are operating normally (as they were this day) to and from an airport, any pilot is going to feel that he too should be able to operate normally. Familiarity with the airport and area might also be an incentive to continue.

If the NTSB's idea that a misidentification of towns might have been made is accepted — and it sounds pretty good to me — then we are back to a simple thing causing a big problem. And we are back to always making doubly sure that any decision is the right one.

On an approach, the margins are down. Where 200 feet off an altitude at cruise is inconsequential, 200 feet too low on an ILS can be a disaster. While 5 miles off en route might not be too bad, on approach it's a huge error. Nowhere is this more true than when operating in mountainous terrain. Here a simple navigational error or a turn in the wrong direction can quickly seal the fate of an airplane. It can't be stressed too much that scud-running is the most lethal form of flying. No matter what you fly and with what equipment, only the precision ILS approach eliminates scud-running. The last part of any nonprecision approach involves a little scud-running; an approach like the one at Sun Valley involves a lot.

MAKE A LIST, CHECK IT TWICE

In the Challenger accident, low clouds in a valley caused the problem. The crew had no way to know in advance exactly what they would see. Sometimes we let far more obvious phenomena bug us. I caught myself on this while shooting a stormy springtime VOR-DME approach at Auburn, Alabama.

The surface winds were moderate-to-strong easterlies and

forecast to stay that way. Auburn didn't have weather reporting at the time, but the reports from Columbus and Montgomery suggested that the weather would be near minimums for the approach. There are actually two VOR approaches, one off Columbus (Georgia) and one off Tuskegee. I chose the latter because it was more nearly aligned with Runway 10 and because the minimum descent altitude is 100 feet lower. I probably felt a little smug about analyzing all factors and choosing the approach that gave me the best deal, even though the approach controller at Columbus had initially started me toward the approach from the east instead of from the southwest.

The air was turbulent at 3,000 feet. The rain was of the showery variety, and as I tracked toward Tuskegee it was apparent that the low-level wind was quite strong. The DME groundspeed showed a gain of about 40 knots with a lot of drift to the right when tracking 320 degrees.

The approach begins from a holding pattern at Tuskegee, right turns 235 outbound and 55 inbound, the latter also being the final approach course. Because a commuter airliner was shooting the approach, I was given holding instructions. I gave a moment of thought on how to enter the hold; the best deal seemed to be to cross the station and then do a teardrop on the holding side. That would mean a left turn to 190 after passing the VOR, flying for forty-five seconds after passing back through the radial, and then turning back to the right.

I was conversing with the controller at station passage and rolled into the left turn as I was talking. It is not a particularly good practice to turn while talking, so I deserved a gig for that. I set 055 up on the VOR, noted that it said "To," checked for a full-scale deflection in what I thought was the proper direction, and then flew for forty-five seconds. Next, a 180 to the right, back around to a heading of 010 to intercept the inbound. My math was flawless. But my navigation was lousy. When I rolled out, the nav needle was full scale to the right and "To" was in the window. Shouldn't it be to the left? The old brain races for a minute when everything is illogical. Finally, I did something logical. I centered the needle to determine the bearing to the station. It was 100 degrees, pretty far from the 055 prescribed for the approach.

What hadn't I done? There had been a complete mental lapse on wind. I knew the wind was very strong and out of the east, yet I made no allowance for it in my convoluted entry. After passing the station, turning to 190, and flying for forty-five seconds I never actually passed through the inbound radial. I was still northwest of it when I turned back to the right to intercept, and on the 010-intercept heading I was actually putting distance between the airplane and the station instead of getting closer as I should have been doing.

At about that time the controller (who never mentioned my lousy navigating in the vicinity of the station) cleared me for the approach. The rest of it was interesting and revealing of the strength of the wind.

The drill was to maintain 2,300 feet until reaching the 9-mile DME fix, then descend to 1,220, the MDA, and fly to the 14-mile fix which was the missed approach point. The going at 2,300 feet was bumpy and oh-so-slow. The airspeed was fluctuating between 130 and 140 in the turbulence, yet the groundspeed was fluctuating between 80 and 85. That was some strong low-level wind.

As I flew on I was heartened that the commuter airliner landed. That suggested that the weather would be well above minimums for my approach because the airliner had come from the other direction. (That might have been a misplaced hope on weather, because the locals later told me that the commuter often made it in when the weather was well below MDA — enough to make me ride the bus instead of the commuter should the occasion arise.) After the commuter landed, he reported severe low-level wind shear on final. That was only logical; the surface wind was 25 with gusts to 35, and the wind less than 2,000 feet above the surface was in the mid-fifties or even a little more. A decreasing headwind means a sinking spell and airspeed decay. I reviewed this in my mind carefully. Having failed to consider wind once, I wasn't about to get caught again.

The runway length was 4,000 feet, and the wind was closely aligned and strong. I thought that I'd make the landing with approach flaps, to keep the drag down and give maximum advantage to the airplane in any tussle with a sinking spell. Target speed would be 100 knots.

The airport was in sight well before I reached the missed approach point, and all that was left was to land — no instrument flight is complete without that successful landing. The airspeed was fluctuating between 90 and 110, so I was happy. When I got to the area of wind shear there was a sinking spell, and it took a shot of power to keep the airspeed from going below 90. And, as is almost always the case, I left the power on a little too long, and the airspeed moved to 110 after I had descended through the shear zone. But the headwind and the runway length took care of the extra speed.

One thing that I kept thinking about during that approach is worth a moment of reflection. The airplane has weather radar, which was showing only light-to-moderate rain with gentle gradients. It also had a Stormscope, which was showing nothing. I was very satisfied that there were no thunderstorms in the area. Yet the ride during approach was rough — quite rough at times. It's funny how we can sit there and feel quite okay about getting banged around as long as the electronic devices say that it's okay. On the other hand we become quite concerned about even mild turbulence when the electronic devices suggest that something bad is around. The latter is good, a form of self-preservation. The good feeling about turbulence when nothing is showing might be charged to human nature, and one should beware this because any turbulence should pique the curiosity and lead the pilot to follow all turbulent-air precautions.

Another thought is related to the commuter that landed before I did. His landing made me think for a moment that everything would be okay for my approach. The Challenger pilots at Sun Valley might have had the same thought. That is a hazardous attitude in IFR flying. It makes no sense to feel that you can do something because another pilot "made it" first. He might have scared the living daylights out of himself and almost crashed. Conditions could get worse between his approach and yours. He might be a more proficient pilot. The list of things is long, and it is far better to make the judgment based on what you observe and what you have at your disposal for the safe conduct of the flight.

NIGHT AND FOG

A lot of factors can bring hazard to an instrument approach; the primary one is the tendency of pilots to leave a safe altitude and descend below minimums when not in visual conditions and without a runway clearly in sight. This is the leading cause of IFR accidents, and it happens more frequently at night than in the daytime despite the fact that most IFR is flown in the daytime.

Why night?

One reason there's more trouble at night is because lights can penetrate cloud and make a false promise. It is as if they are beckoning you to come on down. Many night accidents happen in fog conditions and, remember, in fog the lower you go the less you see.

A friend and I were discussing this one day. The real thrust of the discussion was the question, What would you do if everywhere within range was below minimums? I know that you are not supposed to let that happen. If you do, you made a lot of mistakes. But it is still a worthy discussion, because some unusual combination of blown forecasts and mechanical problems could someday lead to a situation in which the only choice is a bad choice — an approach in very low conditions.

Anyway, the other participant in the discussion thought that he would stick to an ILS approach. He would fly with the autopilot coupled as low as possible, probably as low as 100 feet. Then he would disengage, would do absolutely nothing to the pitch attitude and power setting that had been holding the glideslope, and would keep the localizer needle centered. (The localizer width doesn't narrow to zilch until you reach the far end of the runway, so it's possible to fly right to the ground, in his opinion.) This method would take care of clearing obstacles and getting the airplane to a safe place on the runway as far as altitude goes. And if the localizer needle was kept centered, the airplane should be tracking the centerline of the runway. All that would remain would be touchdown management and rollout tracking.

So much for his theory. I didn't find a lot of holes in it, but said I'd go to a military base and get a ground-controlled approach. It had been a long time since I had flown one, but I always thought they were very precise. All you have to do is fly and follow directions. To boot, a friend of mine made it down in a turboprop twin, with fumes in the tanks, on an emergency zero-zero GCA, and I know of many instances where attempted zero-zero ILS approaches wound up in disaster.

After we had discussed this for a while, the subject of lights came up. Friend said that at night the lights were easier to see, and that if you got over the approach lights, even in zero-zero conditions, you'd know it. And I thought this is where he would kill himself. Those lights would actually offer him a misleading visual cue — unless he wants to land on the approach lights. In such a tight spot I'd ask for the approach lights to be doused; all I'd want would be runway-end identifier strobes, if available, and the runway lights turned up to maximum intensity. That way, the only lights that I would see would be ones defining the place where I wanted to land.

DANGEROUS ILLUSIONS

What any lights do is tempt a pilot to think that the flying has turned visual. Whether on an ILS or a nonprecision approach this can be deceiving. And perhaps the most hazard comes on a circling approach, when the circling is done over an unlighted area.

The woods, fields, and hills are full of airplanes that were left there on nighttime circling approaches. Terrain clearance on a circling approach is assured only by flying at a minimum published altitude while circling within an established distance of the airport. Maybe that sounds easy, but it isn't. What I think happens — and I've seen pilots do it — is that, once the airport is in sight, the pilot fixates on it during the circle. When you fly over a dark area and look at lights in a distance, the visual illusion is of greater-than-actual height, especially if you are looking at the lights through precipitation. The strong tendency is to descend below the published minimum.

A primary item of risk management might be to avoid circling approaches — especially in minimum conditions. If one absolutely must be shot, it should be flown very conservatively.

At night I use the circling minimums for a Category D airplane (ones where 1.3 times the stalling speed equals from 141 to 165 knots) even when flying a light airplane. That expands the margins. (If "D" minimums aren't published, as is the case with some smaller airports, go for "C," which is for airplanes where 1.3 times the stalling speed equals 121 to 140 knots.)

The next way to improve your chances (sounds risky and it is) on a nighttime circling approach is to absolutely and positively remain at the published circling minimum descent altitude until established on final approach to the landing runway. It isn't really until you are on final that you *know* that it's going to be possible to fly visually to the runway. If the circling minimums are 500 or 600 feet, that means a rather long final. That's okay. Obstruction clearance is provided out to the visibility minimum and Category C requires at least a 1½-mile visibility; "D" is usually 2 miles or more. Five hundred or 600 feet can easily be handled over either distance in a light airplane. But if you tried the circle with a mile visibility at 600 feet and had circled a mile from the runway, you'd be closer than a mile when established on final. At ¾ mile, for example, 90 knots means thirty seconds to the runway. From 600 feet, that's 1,200 feet per minute. That is not a normal approach; in fact it would be an almost impossible approach in some airplanes. And that is why the published minimums for circling approaches can be downright dangerous at night. The slower the airplane, the lower the risk, but it is still a risk and is substantial unless quite a bit is added to all the minimum numbers.

A final item relates to the weather conditions encountered during a circling approach. If at any time during the circle a cloud is encountered, that means a missed approach — right then, no kidding. The human-nature response to a cloud encounter when flying visually (which is what we do after breakout on an instrument approach) is often to descend for a better look. That is purely lethal. Safety is up, not down. Published minimums are a safe place to fly; there's no margin for error beneath.

KEEP ALERT

One thing that we often don't think of on an instrument approach is other air traffic. When we are on IFR we tend to withdraw into a cocoon of protected airspace and almost develop a feeling of invulnerability to other traffic. This is something to watch, because there might be VFR flying below the clouds. If one airplane is VFR that means all are VFR when it comes to collision avoidance; being in radar contact and control offers no guarantee.

A business jet on a training flight in a relatively major terminal area was conducting multiple instrument approaches on an IFR clearance. The assigned cruising altitude in the pattern was 2,000 feet. After the completion of an ILS to Runway 17R, the aircraft was cleared to turn right to a heading of 230. Less than a minute later the controller said, ". . . give me a tight right turn now to 350 to pass behind the traffic." This traffic was a single-engine airplane departing to the southwest.

When the jet was clear of that traffic, the tower controller inspected the radar display in the tower and told the flight to turn back to the left to a heading of 320 and contact approach control. This was acknowledged, and the flight contacted approach control. A new heading, 350, was assigned and acknowledged. Shortly afterward, the controller saw the jet start a left turn to 320. The controller said that he was not concerned because training flights ". . . sometimes . . . make turns that they should not make or are not instructed to make." Then the controller saw a primary radar return at the jet's eleven o'clock position at a distance of less than 1 mile. He made no attempt to issue a warning to the jet crew because he believed the two airplanes were at different altitudes.

The other airplane was a single-engine Cessna, inbound to the airport. It was not equipped with a transponder but after contacting approach control was told, "Radar contact, fly heading 120, maintain 2,000. . . ." The approach controller handed the Cessna off to the tower after again telling the pilot to maintain 2,000 feet. In contact with the tower, the Cessna pilot was told to enter a downwind. At this time another tower controller diverted the attention of the controller handling the Cessna back to that airplane by asking if he had traffic west of the airport. The controller saw the Cessna's radar return within a mile of the jet's transponder return. An attempt was made to issue a traffic advisory to the Cessna, but a fireball was seen in the sky before the advisory was finished.

WHOSE FAULT?

As pilots, we might feel comfortable "blaming" that one on the air traffic controllers. The NTSB did, in determining that a probable cause was the failure of controller personnel to separate the aircraft as required by procedures established for a terminal radar service area, to insure that proper coordination was effected, and to issue appropriate traffic advisories. Where we come in as pilots is in the part of the probable cause about the failure of each flightcrew to see and avoid the other aircraft. In studies of cockpit visibility and the relative position of the aircraft to each other, the NTSB determined that the pilots could have seen each other in plenty of time to avoid the collision.

In the case of any IFR aircraft on a training flight, we know how this can be a head-in-cockpit time. The student is usually under the hood if it's instrument training in VFR conditions. The instructor-pilot has to monitor the student and instruct, as well as watch for other traffic. When possible, an observer can be useful on a training flight.

The message is to watch for other traffic when in visual meteorological conditions. Whether in the maneuvering pattern at a radar-equipped airport, or just breaking out of the clouds at 800 feet on a VFR approach to an out-of-the-way airport, the responsibility to see and avoid VFR traffic becomes operational whenever the IFR airplane is in VFR conditions. And while it is important all the time, it is most critical on an approach, in the vicinity of an airport.

When flying IFR to my home airport in New Jersey, the arrival is handled by Philadelphia approach control. One of their procedures is for all airplanes to descend to 5,000 and then to 3,000 a considerable distance away from Mercer County Airport. This puts the flying under a fold of fat on the Philly TCA. I was always alert for other traffic during this phase of flight; then one day when I was returning from a memorable trip to 51,000 feet in a Learjet, another pilot added emphasis. The keeper of the Learjet, Bill Benke, a former Air Force pilot, seemed amazed that in a Learjet we were so low so far from the airport and were mixing with the VFR traffic compressed under the TCA. He told me to raise the seat, get some flaps out, slow the jet to give us more time to see and avoid other traffic, and fly heads-up. Sure we were IFR, and sure we were in radar contact, but this wise aviator of many campaigns knew that the art of collision avoidance was up to us, and the conditions in which we were flying were ideal for the practice of that fine art.

BRAINY BUSINESS

Whether the problem is flying into another airplane, flying into the ground, or simply losing control of the aircraft, approach problems can usually be linked to the pilot looking at, thinking about, or doing the wrong thing at the wrong time. We know that the pace quickens as the approach progresses; how do we troubleshoot our performance and refresh ourselves on what to do and what to think about during the approach to keep it from literally becoming a "final" approach?

The first evaluation backs up to the arrival phase of flight. When determining that an approach did or didn't get off to a good start, we are grading the arrival. Was the airplane in the desired configuration and was it flying at a proper speed and altitude on reaching the final approach fix? If no was the answer, bad grade. The approach probably reflected the quality of the arrival performance.

Was the pilot briefed on the approach before reaching the final approach fix? When solo, we brief ourselves. Nothing wrong with that. Do it methodically.

A friend of mine will forgive me, I hope, for revealing his lapse in an approach briefing. It was a low day, it was a nonprecision approach, and it was bumpy. I've never flown with him, but would think he is a methodical sort in an airplane. But he sure missed one item on his approach briefing. He didn't tell his wife or one of his children to hold onto the dog. When the bumps started the dog didn't like it, and as the pilot was descending to the MDA he wound up with a dog in his lap. Needless to say, he missed the approach. (Serves him right, too.)

The grading of the approach itself begins at the final approach fix, with the airplane inbound to the airport.

Was the aircraft promptly established in a descent that would take it to the missed approach point at a proper airspeed and in a proper configuration? Was the heading one that would be expected to track the final approach course? Was aircraft control and the approach itself put ahead of communications requirements? Were the details of the approach in mind so excessive time wasn't devoted to the chart after the approach started? (I like to write the MDA or DH and the first of the missed approach procedures on a piece of sticky paper and put it on the panel.) The answer to all those questions should be yes for a good grade.

Once established, was the tracking of the approach smooth? Were corrections well thought out?

The mind must be very active when flying an approach. To keep up with a changing wind effect on the approach, we have to evaluate continually the relationship between the heading flown and the movement of the navigation needles. It is the act of flying a constant heading and evaluating the effect of that heading that allows us to make intelligent corrections rather than lunges at the final approach course. If we don't settle down, fly one heading, and determine what that does, the best we'll do is *S*-turn down the final approach course.

The descent works the same way, especially if a glideslope is involved. We know a power setting and attitude that yields a correct airspeed while tracking an average glideslope. Then we make adjustments for groundspeed changes resulting from changing wind, or for glideslope facilities that are a touch shallower or steeper than normal. Just as with heading, the drill is to identify the need for a correction, make the attitude or power correction, and then evaluate the effectiveness of that correction. If the glideslope is chased, it's a vertical-*S* maneuver above and below.

A key to smooth flying in this phase of the approach is in looking at the correct thing at the correct time. Usually, when a pilot is having approach problems, a little study will show that he is looking at the wrong thing at the correct time.

For example, if a pilot is doing *S*-turns and weaving back and forth through the final approach course, odds are that he is looking at the navigation needles when making corrections. This doesn't even work when the nav is combined with the heading, as in a horizontal situation indicator, because you fixate on the nav part of that instrument when making a turn. It works even less well when the information is scattered all over the panel. The way to make a correction is to observe the present heading, observe the position of the nav needle, determine a new heading that will move it in the proper direction, and then turn to that new heading using the attitude and heading indicators. Just thinking, "I need to go to the right" and rolling into a right turn doesn't work.

The same goes for corrections on the approach slope. If everything is steady but the airplane is a little high on the glideslope, or if it appears that the minimum descent altitude won't be reached before reaching the missed approach point, then we know that the rate of descent needs to be increased. If the decision is to reduce manifold pressure by two inches, stay with the flight instruments and lower the nose slightly while moving the throttle back just a little. If you don't know the airplane well enough to know how much throttle movement results in a two-inch decrease in manifold pressure, you need basic practice. After the power adjustment is made, glance at the power instruments to verify that the change was about as desired. Then refer to the glideslope or to the clock and the altimeter to ascertain that the power change is having the desired effect.

On any approach, an important part of the descent is keeping track of altitude. This is easier if we act as a crew of one and call out to ourselves, when we're at 1,000 feet, 500 feet, and 100 feet above the MDA or DH.

SMOOTH APPROACH

One of the smoothest airline rides that I ever had was on an Air Canada flight, into New York's LaGuardia airport. After the aircraft left cruising altitude to touchdown, only two power adjustments perceptible in the cabin were made. The descent was with the aircraft clean until we were close to the marker. The first power change came when the captain configured the aircraft for the final approach — gear and flaps down. It was a 727, and the power is increased markedly at this time. The next power change came in the flare. The pilot had a perfect feel for the dynamics of his airplane, and he set the power so close to right the first time that any required adjustments were minor enough that you couldn't hear or feel them. That's what we should all strive for in an approach. When we wind up horsing around with an airplane on the approach it is a sign of alternating mental paralysis and bursts of power if not brilliance.

As any approach progresses, we have to make an honest judgment of performance. Is the work that is being done good enough to justify a continuation of the approach? That's something to think about all the way in. A sure way to jangle yourself is to persist in trying to salvage a poorly executed approach rather than saying, "Hey, self, this didn't get off to a good start. Miss it early, get things together, and do it right." If there is some underlying factor that caused the approach to get fouled up, such as turbulence or wind shear or nervousness over near-minimum conditions, perhaps it's best to go to Plan B rather than to persist.

When all is said and done, most approaches work well. Not all are perfect, but we generally manage to hold things within tolerances and earn a view of the runway. When it doesn't work, the next step, and the next chapter in this book, is the missed approach.

CHAPTER 7

The Missed Approach

FIRST PART Patrick E. Bradley

Through the course of my instrument training, I must have flown hundreds of missed approaches under the hood. My instructor's routine was to request a low approach, have me shoot the approach, execute the missed approach, and try it again. During these exercises, my primary concern was remembering the procedure. At the missed approach point or decision height, I would apply full power, pitch up, reduce flaps one notch, and when I had a positive rate of climb, retract the gear. At first, I was like a bull in a china shop. If I got the power in, flaps up, and gear retracted, I would forget to pitch the airplane up. Then I'd forget the missed approach procedure. I'd end up asking my instructor, who had long since taken the approach plate from me. At other times I would remember the procedure, but forget to raise the gear or the flaps.

INSTRUCTOR TO ME: Not climbing too well today, are we?
ME: Ah, no, but it's hot today, that might be the reason.
INSTRUCTOR: I think there might be another reason.
ME: Forgot again, didn't I.
INSTRUCTOR: Yup.

In time, with lots of practice, I got pretty good at missed approaches; the procedure became second nature. I was confident that when the time came, I wouldn't have any problem with the dreaded missed approach. But I had another think coming, and I'm thankful that Collins — who has missed approaches before — was with me when my time came.

On a return flight from Washington National to Trenton, New Jersey, I was ready for some exciting weather. National was 500 overcast with a mile visibility, and Trenton was forecast to be slightly worse. In fact, Trenton was quite a bit worse. First I missed an NDB approach at Doylestown, near Trenton, for "practice." Then as I began the ILS to Runway 6, the weather was reported 200 feet overcast with ½-mile visibility. As it turns out, that was an optimistic assessment, but for some reason I never thought for a moment, even with the dismal forecast, that I would have to miss the approach. I had never not completed an actual approach before this day, and down deep I just didn't think that missed ILS approaches happened outside of the training context.

Although my ILS at Trenton had not been perfect, it was acceptable, and reaching decision height, I had every intention of looking up to see the sequenced strobes of the approach light system. Instead, I looked up to see a blank wall of fog. No lights, no nothing. I looked around, and then straight down, where I did see lights. I'm still not sure whether they were the runway lights. But they were clear enough to throw me into a state of confusion. I had dropped 20 or 30 feet below decision height, and was still flying along without making what seems the obvious decision to get out of there. Luckily, Dick made the decision. With some prodding I began the missed approach, but was stricken with the old bull-in-a-china-shop syndrome once again. I got the full power and flaps part, but I never pitched the airplane up. This time, I had forgotten the most important element of a missed approach. *Pitch up*. We tried another, it was worse, and we finally got out of there, ending up landing in Allentown, Pennsylvania.

Missing that ILS was one of the most valuable lessons of my instrument flying career. First, I learned that missed approaches do really happen. Although they may come few and far between, every active instrument pilot will, at one time or another, have to miss an approach because of below minimums weather. In retrospect, I see that preparing for a missed approach mentally is just as important as learning and practicing the procedures. Without realizing that yes, this is an approach that I may very well have to miss, it becomes nearly impossible to follow through with the miss quickly and deliberately enough to remain safe. Second, I learned that the decision to miss an approach just isn't the sort of proposition that I could mull over and weigh when I arrived at and continued beyond the missed approach point. Deciding whether to continue or miss an approach is an immediate act. I was confused by the lights below the airplane. I wasn't quite sure whether I saw enough of the runway environment to land the airplane. Now, when I reach DH or the missed approach point, I either see the runway in front of the airplane, or I immediately miss the approach. For my own safety, I made the policy decision that unless I have a clear view of the airport and runway, I'm getting out of there in a hurry. For me, there just isn't enough time to weigh options; it's a cut-and-dried decision. Even the least doubt is sufficient reason for me to miss the approach.

Missed approaches in real-life instrument conditions are as different from practice missed approaches as night is from day. I never realized the advantage I had in my training sessions, knowing beforehand that I would miss the approach. I had the decision made for me even before I began the approach. In actual conditions, the line between an approach that a pilot can complete and an approach that he is obliged to miss might not be apparent. The only alternative is to impose a clear distinction. Either the threshold of the runway is there — without a doubt — or it's time to miss. Of course, even this is easier said than done. Often, a missed approach will mean continuing the flight to another airport, renting a car, and driving to your destination. Overcoming the desire to get home can require great discipline. But that's another reason for setting clear guidelines as to when to miss an approach, holding them above all other concerns: It's really the only way to be certain you can fly a close approach safely.

I consider myself lucky to have been able to experience an actual missed approach with a seasoned pilot along. I learned how simple it can be to fall into the trap where, although you know how to miss an approach, you really don't have a clue as to what it really looks like and feels like to do so. About three months after my first missed approach at Trenton, I "missed" another approach in actual conditions. Before beginning this approach, I learned that the ceiling at Trenton was nonexistent, and that several other airplanes had missed the approach. That was enough for me to forgo the approach altogether. I diverted to Philadelphia International. That is the ideal missed approach, but less than ideal situations are bound to arise, times when it's not clear whether the conditions will be good enough to complete the approach. At times like that, I think back to my first missed approach and tell myself: "This approach may or may not work out. If I have any doubts when I reach decision height, I will have no doubts as to my next step." I will happily miss and try again another day.

SECOND PART   Richard L. Collins

A missed approach in actual conditions is one of the tougher challenges in instrument flying, and there are a lot of serious accidents in this phase of flight. The difficulty of the task is enhanced by three things. First, no pilot expects to miss an approach, nor does any pilot want to miss an approach. The desire simply isn't there. We want to land, or we wouldn't be shooting the approach in the first place. Second, the missed approach is, from a technique standpoint, demanding. About the only thing that doesn't change is the airspeed. There's a power change, airplane configuration changes (unless you are flying a very simple airplane), and trim changes. All this is done at a low altitude, on instruments but perhaps with the distraction of things moving by in the murk and gloom close below. And I do mean close. On the ILS approach at my home airport, Mercer County, New Jersey, the decision height on the ILS is 386 feet. That's 200 feet above the touchdown zone on Runway 6, which happens to be the lowest point on the airport. Slightly left of the runway, on higher ground, is the control tower. It tops out at 271 feet, or, only 115 feet below the decision height for the runway. If you miss and are a little to the left and a little slow to get the airplane climbing, it's no wonder the tower calls and says, "I hear that you just missed the approach." Some years ago the tower crew found itself eye to eye with a pilot who decided late, low, and left to miss the approach. The tower was subsequently painted in a bright contrasting red and white. Memories fade, though, and they recently painted it a soft tan, to blend in with the background.

I'll bet that they paint it red and white again, someday, in honor of the missed approach.

The third reason a missed approach can be difficult relates to hardware. If we have a flight director/autopilot system and if we let the autopilot shoot a coupled approach, pressing the "go-around" button disconnects the autopilot and commands a go-around attitude on the flight director. The pilot has to, at one of the most difficult times imaginable, revert to hand flying. If the pilot is sharp, that's okay. If the pilot isn't sharp, it's something else again.

I got a good refresher on missed approaches watching Bradley miss his first actual approaches because the weather was below minimums. You've read his version, here's mine:

FOR REAL

The flight was from Trenton, New Jersey, to Washington National and return. The weather was low when we left Trenton, 200 and ½, but was forecast to lift to 500 or 600 feet and a couple of miles visibility. We were filed back to Doylestown, near Trenton, because I wanted to let Patrick have an extra approach, an NDB there plus the ILS into Trenton.

As we flew back toward Trenton from Washington, the ATIS told us that the NDB to Doylestown would likely be a miss. Trenton was still reporting 200 overcast and a ½ mile visibility.

The missed approach at Doylestown was successful but without gusto. Patrick knew he was going to miss, but there wasn't an enthusiastic entry into the missed approach procedure at the magic moment. I should have taken that as a clue that training does not get pilots ready for actual missed approaches. But Trenton was reporting minimums so we could make it in there, couldn't we? The weather is usually better than reported by the tower because they are on higher ground than the approach end of the runway.

I said not a word to psyche Patrick up for another possible missed approach. He flew the approach reasonably well, but at the magic 386 feet there was nothing. The needles were about crossed; he said he thought he saw a glimpse of the approach lights, but I didn't see anything ahead and I don't look down. Bradley did a pretty good job of leveling off at 386 feet, but that was all. He glanced at me with that "what next?" look and, if I remember correctly, I told him that he'd better get the airplane into the missed approach procedure and configuration or he'd bust our tail. With that, things picked up. The tower heard us go over.

It is not particularly good practice to shoot an approach again after failing to see something when at the decision height unless there is evidence that the weather has changed. If an approach is abandoned early, that's another matter. But if you can't see at 200 feet and nothing has changed, going again is looking for trouble. There are a lot of accidents on second approaches, when pilots try harder to convince themselves that they can indeed see something. But I wanted Patrick to have another shot — not so much at the approach but at the actual missed approach. They were still reporting 200 and ½ so we came around for another one.

This time he never got the approach stabilized. His thinking must have moved forward to the missed approach early, while he was flying the ILS. Then he made a tactical error, which was good for him to see. He strayed above the glideslope about 200 feet above the decision height, and when he saw the needle deflection, he went after it. The correction was too much, and he went through the glideslope at about the decision height.

The force he felt pulling on the wheel was coming from the right seat. I transitioned the airplane to the missed approach, mainly because I thought he might not understand the immediacy of the moment. We had gone a bit below the decision height, and I wanted it back up as quickly as possible.

I was going to give him a third shot, but they changed the weather report to 100 and ¼. It was time to go sniffing for an alternate.

En route to the alternate, we discussed the approaches as well as the day. There were airplanes all over the place, missing approaches and fleeing to alternates. The Navy Reserve had launched its fleet (the day was Sunday), expecting reasonable weather for recovery at Willow Grove. But the weather was below minimums there too. I remarked that I hoped nobody would try too hard; sadly, they did. A 310 crashed a bit later, after calling missed approach at Atlantic City.

INTO THE WEEDS

An accident involving a T-1020 (a commuter version of the Chieftain) offers some further points to contemplate on the missed approach.

The aircraft was approaching Durango, Colorado. It was dark, and the reported weather was indefinite ceiling 400 overcast, with 1-mile visibility in light snow and fog. The airport was served by a VOR approach, with a minimum descent altitude 333 feet above the ground. The first officer of a Frontier flight which landed at Durango shortly before the T-1020 arrived stated that cloud tops in the area were about 10,000 feet with bases near 7,000 feet. (Field elevation is 6,685; the MDA is 6,980.) The first officer further stated that their airplane picked up a trace of light, to at worst moderate, icing in clouds and snow. He said there were no significant wind shears, gusts, or turbulence during the approach to Durango. He said that, once on the ground at Durango, he had estimated the visibility at 1 to 2 miles. Eyewitness accounts and a slightly later official observation indicated a deteriorating ceiling and visibility at the time of the T-1020's approach. The measured ceiling at the airport dropped from 400 to 200 feet and the visibility from 1 mile to ½ mile at and shortly after the approach.

The aircraft was seen by several witnesses as it missed the approach.

One witness said that his attention was drawn to the aircraft because of the engine noise and the fact that it was very low. He said it flew parallel to Runway 2 until it disappeared into the snowfall. He said the wings were level. Another witness saw the airplane fly along the runway, make a right turn to 060 (which is the missed approach heading), and then descend behind a tree-covered hill. He estimated the airplane's altitude as it passed to be about 400 feet. A third witness said the airplane was about 200 feet high and 200 feet left of the runway as it flew by. He also observed the turn to a heading of 060. A fourth witness said the airplane flew over the runway at an altitude of about 100 feet. The first officer in the Frontier flight on the ground at Durango said he heard the flight approach the airport and observed it make a missed approach. He said that shortly after he observed the right turn, the airplane attitude abruptly changed from level flight to a sharp descent instead of the climb that he expected.

The airplane first contacted sagebrush and then a fence on a level, snow-covered pasture about 1,350 feet east of the departure end of Runway 2, at an elevation of 6,592 feet. The airplane stopped 1,600 feet farther along. The crash path was aligned 060, the initial missed approach heading. The landing gear and flaps were retracted, and there was no evidence of any preimpact failure or pilot medical problems. The engines were apparently developing high power at impact. In its report on this accident, the National Transportation Safety Board postulated the required rate of descent from the missed approach point to the crash site. If the airplane had been at 300 feet over the airport, the rate of descent would have been 1,200 feet per minute; from 200 feet it would have been 800 feet per minute, and from 100 feet, 400 feet per minute. The evidence indicated that the pilot probably recognized the unwanted descent at the last moment and tried to climb. The initial ground contact primarily involved the props cutting through sagebrush until the aircraft passed through a fence a few seconds later. Then the aircraft remained in the air for another 450 feet before hitting the ground.

The NTSB was not able to determine the probable cause of the accident, but they did delve into all the possibilities.

POSSIBILITIES

The airplane had an electric trim system; the Board noted that the trim could have run away concurrently with the airplane configuration changes accompanying the missed approach. However, there is no history of runaway trim problems in this type airplane, and even if it did run away, design criteria specify that the pilot must be able to overcome the forces involved with runaway trim.

The airline's unwritten procedures suggest that pilots not use the autopilot for approaches, but the possibility exists that this pilot did use it for the approach to Durango. A survivor reported hearing a "bell" as the airplane entered the missed approach, but there are no bell warning systems in this type airplane. The sound of the autopilot disconnect is a "bong," and that might have been the sound heard by the survivor. If it was, it could mean that the pilot was using the autopilot in the approach and disconnected it for the missed approach. If the autopilot had been on, had malfunctioned, and initiated the descent, or the pilot made an improper input, the autopilot could easily have been turned off.

The aircraft was equipped with approved deicing gear, and based on the Frontier pilot's report and existing conditions, the Board decided that ice wasn't likely a factor.

The Board considered the possibility that the pilot was executing an intentional maneuver, an attempt to remain visual after seeing the airport go by and to circle to land on Runway 20. (The wind was calm.) However, the pilot was well aware of the surrounding terrain features and would hardly be expected to attempt such a maneuver by losing altitude.

A failure of the attitude indicator was considered, but lacking evidence that it did, and with the airplane flown normally to the point of the missed approach, the Board decided that the attitude indicator probably wasn't a factor.

The pilot's work load in the missed approach was analyzed, and in reference to the few seconds after the missed approach was begun, the Board stated: "The work load on the pilot would have been extremely high during this short period and would have required precise and timely perception and decision making, and actions to effect a proper missed approach."

The possibility of spatial disorientation was examined. ". . . the possibility . . . is greater at night, during aircraft configuration changes, and when accelerations (aircraft maneuvering) occur, i.e., turns, climbs, and power changes." Also a passenger screamed something about another airplane on the runway in reaction to the start of the missed approach. There was no other airplane on the runway, but the scream could have caused the pilot to look down and to the left, and head and eye movements can induce spatial disorientation.

The Board also flogged the single-pilot horse one more time. It's interesting how when an airplane crashes with two or three pilots up front, they lay no blame on the larger number in the cockpit even though at times the interface between pilots appears to have a relationship to an accident. But when there's only one up front, the lack of a partner is always suspect.

LESSONS

There are a lot of lessons in this particular missed-approach accident as well as in the experiences related with missed approaches.

The clearest message is that the missed approach needs some thought in advance, at a quiet time before the approach begins. We need to carefully think through the things that will make us miss an approach and how we'll handle each one.

Weather is a primary factor in missed approaches, but I think the NTSB was entirely incorrect when it added, after saying it was unable to determine the probable cause of the T-1020 accident, "a low ceiling and poor visibility were factors which contributed to the accident." Those things are as accepted in instrument flying as is the air in which to fly. The total of the efforts of the system and pilots is directed at keeping low ceilings and poor visibility from being a factor in accidents. Our part of the job is to evaluate the likely effect of weather on the approach and then to be ready for *anything*.

I almost missed an approach at Indianapolis one day when the automatic terminal information service had informed me of a good ceiling and visibility. A heavy snow shower blew over the field, obliterating everything, and the tower controller never mentioned it. When I complained about not being advised of the deteriorating weather while taxiing in, the controller said, "If you can't stand the heat, stay out of the kitchen." He was right. I taxied quietly the rest of the way. I knew there were snow showers around and should have thought through the possibility of encountering one on final before beginning my approach. On any approach the possibility of a miss or a go-around should be cheerfully accepted.

EASIER?

I think there is much less sweat to missing a nonprecision approach than missing a precision one. The nomenclature is evidence of the differences in the two approaches. On a nonprecision you let down to an altitude and fly for a time or a distance. On a precision, you have to pat head and rub tummy simultaneously. Keeping it on a rather sensitive left/right as well as on a more sensitive up/down requires a faster thought process than flying a not-so-sensitive left/right and a reasonable rate of descent to a predetermined altitude. The pressure is even throughout a nonprecision; it peaks right at the end of a precision approach. On a nonprecision you transition from level flight to a climb; on a precision it's from a descent to a climb. There is more to the latter. As a result of all this, initiating a nonprecision approach when there's a likelihood of a missed approach because of weather is a better deal than initiating a precision approach when that likelihood exists. Two hundred and $1/2$ isn't much, and as for Category II (basically 100 and $1/4$), for me, no thanks.

The decision height is clearly defined, and if the runway isn't in view ahead at that point on a precision approach, the airplane should bounce off the minimum altitude like a Ping-Pong ball. If it's a nonprecision, the airplane should start up at the missed approach point.

If there's any question about the weather, think through the drill in advance. And in using the callouts discussed in the previous chapter, take the "100-foot to go" call as a signal to cock the go-around pistol. At that point the burden of proof is on the pilot and if the proof isn't there, in spades, at the correct time, boom, go. No questions, no hesitation.

In the case of the commuter airline T-1020, there's probably little question that the pilot saw the runway lights as he flew over. But there's also no question that it was too late to land when he saw them. The airplane was too high, and it might not have been quite aligned with the runway. On most approaches the pilot is somewhat on his own here. On an approach with a visual descent point specified, we know that passing that point without the runway in sight means that a later sighting can't lead to a normal descent to the runway. The lack of a VDP on an approach plate doesn't mean we can't calculate one of our own. The approach at Durango is a VOR/DME, the vortac is on the airport, and the missed approach point is shown as the vortac. In other words, according to the approach plate, you fly to a point right over the airport before missing. But it's easy to know that you've missed a straight-in when the DME has counted down to 1 mile and the airport isn't in sight dead ahead. On approaches that use timing as opposed to DME the missed approach point is usually the end of the runway. But, in reality, a straight-in approach is as good as missed if you don't have the runway in sight a mile before reaching that missed approach point — higher if you are more than 500 or 600 feet above touchdown zone elevation. All you have to do is know the time when you've likely flown past the point from which it would not be possible to make a normal descent to the runway from the final approach course.

CIRCLING MISS TO OBLIVION

Another accident expands on the subject and introduces the hazards of trying to convert a straight-in nonprecision approach to a circling approach instead of a missed approach should the runway become visible after it's too late for a straight-in.

There was no weather reporting at the airport as the pilot approached for a twilight VOR approach. An airport 14 miles away was reporting 300 overcast and a mile visibility, though, so the pilot should have been aware of the strong possibility of a missed approach, because the minimum descent altitude for the VOR approach would leave the airplane 600 feet above the ground, or, 300 feet above the cloud bases reported nearby.

The controller vectored the airplane to the final approach course, and when the airplane was 12 miles from the airport, the controller cleared the pilot for a straight-in approach. The final approach fix was defined by DME and the missed approach point by timing from the final approach fix.

The controller continued to observe the aircraft on his scope after issuing the approach clearance and approving a frequency change. He watched the target continue inbound and observed the altitude readout continue down to the 800 foot MDA, where the readout was lost. He continued to monitor the primary target and observed one more readout of 400 feet, which, if accurate, would have been only a couple of hundred feet above the ground. The target crossed over the airport, turned from its southerly heading to northerly, and then passed the airport from northeast to southwest. The target then turned back northeast and disappeared from the radarscope.

Several witnesses at and near the airport saw or heard the airplane. They all described the airplane as crossing Runway 16 from the west side to the east. Those who saw it described it to be about 150 to 200 feet above the ground and in and out of clouds. One witness who saw it fly across the airport said that the landing gear was down and the flaps were partially extended. He described the airplane as circling to the east of the airport with the engines at high power and the props out of synch. Other witnesses near the south end of the airport saw the airplane moving in and out of clouds heading south-west just above the trees. Some described the airplane as climbing and descending in and out of clouds while others said it was level but appeared and disappeared in the ragged overcast. All agreed that it was just above the tops of the trees. Witnesses watched it turn abruptly right and then saw a flash near the right wingtip. The airplane then rolled sharply right and went straight down into the trees.

There's not much question that the weather was probably below minimums for this approach. Yet the pilot cleanly got away with an initial decision to descend below the MDA in search of the airport. Then he moved into what is clearly one of the most hazardous forms of flying: Instead of going into the missed approach he attempted a circling approach below minimums. It happens that the straight-in and circling MDA is the same for this approach — both altitudes well above that at which he was flying — but once a straight-in approach is begun it is probably a questionable practice to convert it to a circling approach — especially in twilight or night conditions.

ABOVE AND BEYOND

The question in this case goes far beyond normal operating practices and the regulations. What does it take to dictate a missed approach after all caution has been thrown to the winds and the haven of a safe altitude has been abandoned for what is nothing more than the worst form of scud-running? This pilot had a lot of chances, even after making some bad decisions. But he apparently tried to circle and line up with the runway instead of embracing that published missed approach procedure that would have saved his and his passengers lives.

The fact that an airport has no weather reporting shouldn't affect the safety of a flight or any of the pilot's decisions on an approach or a missed approach. It's strictly a procedural matter to descend to published minimums and then make the transition to the missed approach procedure if the runway isn't in sight and a normal approach isn't possible. True, missed approaches aren't the easiest maneuver in the world, but properly done by a proficient pilot, they are perfectly safe.

There are often reasons to miss an approach before reaching the minimum altitude for that approach.

Meteorological conditions other than ceiling and visibility might dictate a miss. For example, the runway or conditions associated with it might be something you would rather leave alone. You might reach that point in space from which it wouldn't be possible to complete the approach even if the runway were to come visible. That's a sure time to miss.

BAD SCENE

One of the worst temptations comes when the runway is sort of in sight. This happens most often on a circling approach, but it can also happen on a straight-in. It's more likely at night, and it involves scud, shallow-ground fog, or both.

The runway, or the runway lights, might indeed be visible from some angles. Flying over the airport, the pilot might be able to look straight down and see the runway or lights rather clearly. That generates pressure to land. But, especially in the case of fog, we have to remember that the lower we go, the less we see. In fog, vertical visibility can be a lot better than horizontal. The missed approach in this case needs to be started early lest it have to be started from a very bad location.

I was approaching an airport in conditions that could only be described as good at first sight. The vector was for a straight-in ILS, and the runway lights were clearly visible from the outer marker. Some shallow ground fog was also visible, but it looked like a wispy nuisance. Carry on. The runway remained visible all the way in. At the decision height there was no question that the lights and the runway were in sight, and the airplane was in a position from which a normal landing could be made. At 100 feet it was still true. Then at 50 feet, zap, the visibility went to zilch. What to do? The rule said missed approach, but by the time I got started on that I realized that I was about to touch and was aligned with a stripe on the runway. So I landed. Under the circumstances it seemed a better deal than initiating a missed approach from under 50 feet. On the rollout, I rolled out of the wispy fog and into brilliantly clear conditions. It was a good illustration of how you see less the lower you go when there is fog. This is true whether it's shallow ground fog, or a real full-blown fog.

DUCK

Another common "now you see it, now you don't" involves scud or a combination of scud and fog. Here we see the runway on a "sort of" basis. It's dimly visible around the corner of one piece of scud or another and then it disappears. A duck under, to get beneath the scud, is totally hazardous. Scud is often based on the ground; once you leave the MDA you are strictly VFR, and judging height is all but impossible when you are busting in and out of wispy clouds. There's just not much going for such an approach. The trick is to stay at the minimum descent altitude, and if the runway is not clearly visible from there, miss the approach.

It would be nice if all ceilings were clear-cut and all visibilities were exactly as reported. Unfortunately they aren't; it is the pilot's job to make the good decisions and miss the approach from a safe altitude if the runway isn't in the right place at the right time.

SHEAR

Wind-shear turbulence on final is another reason to consider a miss, especially in a heavier airplane headed for a minimum length runway. Any pilot worth his salt is going to build in a little airspeed pad if it is suspected that wind shear will cause an airspeed decay on final. It's okay to pursue this for a while, but the pilot, we'll hope, establishes a point on final where an unstabilized condition means a missed approach. And at busy airports, where the traffic is continuous, an airplane ahead that misses should arouse total suspicion of conditions. Lightly passing off another pilot's report of wind shear on final as the reason for a missed approach can lead to a tussle with the approach lights — and the airplane always loses.

In light airplanes, the wind-shear problem isn't as bad as in heavy ones because we can accelerate more rapidly and better handle sinking spells and decays in airspeed. We can also fly approaches grossly fast and still land on the relatively long runways often used for instrument approaches.

An unstabilized approach in a light airplane is more often related to excursions on the localizer and glideslope than to speed and configuration. There's a point where such excursions have to be allowed to dictate a missed approach.

A good example comes in the missed ILS approaches at Mercer County discussed earlier in this chapter.

On the first approach, Patrick did a good job of keeping the needles centered to the decision height. No complaints there. On the second approach, he never did get things tied down. He was thinking about something else, and the approach was loose and sloppy. He later said that he started to abandon the approach early, and to try to psyche himself up in the go-around for a fresh start. Indeed, that was what he should have done, but he still could have avoided my having to provide control inputs to arrest a descent below the decision height without the runway in sight. The key is in having a preconditioned reflex *not* to do certain things. In this case, it would be not to increase the rate of descent to recapture a wayward glideslope when the airplane is below, say, 400 feet above the ground. If that might be a good rule for light airplanes, a rule for a heavy one would be higher because the heavier the airplane, the more difficult it is to arrest a high rate of descent. In reality, Patrick had reached his practical minimum for the approach when he effectively lost the glideslope about 400 feet above the ground. A recapture from that point would be unlikely, and the consequences of diving for the glideslope would be far too chancy.

There's also a time to let localizer excursions dictate a missed approach, and to me, this relates more to needle movement than to needle deflection (within reason). If the airplane (again referring to light airplanes) is below 500 feet above the ground on an ILS and the localizer needle is moving rapidly in either direction, that is an indication that the pilot's control of heading is in serious question. Rather than mount a salvage job, the best thing is probably to forget the navigation needles, go to attitude instrument flying, the neglect of which was probably the cause of the rapidly moving needle, and establish the airplane in the missed approach configuration on the proper heading.

THUNDERSTORMS

There are times when thunderstorms must be allowed to dictate a missed or abandoned approach. If the thinking processes were really intact, the approach wouldn't be started if thunderstorms were a factor along the approach course or near the airport, but anyone can be guilty of the foolish optimism that gets an airplane pointed toward a bad place at a worse time.

It's easy, in armchair flying, to talk about how an airport will automatically be rejected for use if there is a thunderstorm nearby. But in reality, we occasionally take a crack at an airport when thunderstorms are threatening. With the goal so near and the airplane operating in good conditions, it's difficult to acknowledge the possible existence of an impossible maelstrom along the way. But if there is such, the earlier it is acknowledged and the earlier you go away, the better. I recall being on an airliner one day when the captain rejected an approach when he was inside the marker because of a "little ol' cell by the airport." I just naturally wondered what took him so long. The margins in the last part of an approach, and in a landing, are surely not adequate for fooling with the mechanics of a thunderstorm. And the ultimate bad trip would be where you get to minimums, realize that it isn't going to work, and then have a nice bubbly thunderstorm in which to conduct the missed approach.

OTHER REASONS

As you fly instruments there will be occasions for a missed or abandoned approach for equipment-related reasons. If there is doubt about equipment, best sort that out sooner rather than later. Taking an airplane with suspect equipment to an airport with minimum weather just isn't a brilliant thing to do.

Wake turbulence might be another reason to miss an approach. We know that wake turbulence settles, and this is generally construed to mean that we can follow a heavy airplane down a glideslope and have its wake remain below. (I'm chicken and always stay a dot high on the glideslope when chasing a heavy, just in case he deviates a little.) But what if you are following a heavier airplane, and it misses the approach? You better know the answer if it ever happens because if you don't follow suit and go around as well, you'll have to fly through its wake to get to the runway.

Whatever the reason, once the decision is made to miss an approach, the flying has to be good and the thinking methodical. If the procedure calls for a turn, the turn has to be made in the proper direction. I know of at least one case where it was in the wrong direction and into the waiting arms of a mountain. And, as is true in other phases of instrument flying, the airplane has to be properly flown. We saw where distractions play a role in loss of control during a departure. A missed approach can be even more distracting because it is an involuntary departure, a demanding time to fly, and an event that can make the mind race to thoughts unrelated to the proper conduct of the immediate task.

CHAPTER 8

When Things Go Wrong

FIRST PART Patrick E. Bradley

Lurking behind the scenes in every flight is the possibility of some sort of malfunction, be it electrical or mechanical. But malfunctions, for me at least, hardly ever happen. Maintaining "combat-ready" status with regard to glitches can be very difficult in the course of day-to-day flying. No one wants to dwell on the fact that machines, like humans, aren't perfect. As a safe pilot, though, I know that I have a duty, to myself and to my passengers, to be prepared to handle every emergency that may rear its ugly head. That requires a reasonable preoccupation with the possibility of an emergency. I call my personal program "functional preparedness for technical malfunctions," or "staying current on glitches." First, I learned the emergency procedures. This step took place during my private and instrument training. It was book work mostly, and consisted of going through the airplane manual and learning the steps manufacturers recommend for any given problem. Of course, I've got to repeat this step for every airplane that I'm checked out in. Different airplanes have different emergency procedures, even simple singles.

Second, I practice the procedures. Without practice, it's not really possible to prime yourself for the real thing. My review is often in the airplane, but I also try to stay abreast of emergency procedures in simulators. They come closest to reproducing the surprise element that is, to some extent, part of every emergency. Finally, I try not to delude myself that malfunctions are not a part of flying, no matter how infrequently they occur. I find this the most difficult aspect of preparing for emergencies — accepting that they happen, and that no matter what, I can deal with them.

Studying recommended emergency procedures is important for a number of reasons, one of which is to learn which steps to take before referring to the checklist, and which steps to leave until the checklist is in hand. I've found that even if I know all of the steps for an emergency, it's always important to run through the checklist. I first learned this lesson when I was checking out in a Beech Sierra. One of the first malfunctions my instructor pulled was an emergency gear extension. He wanted me to describe the proper procedure. I was ready for this, and carefully checked to see that the master switch was on and that the gear lights weren't turned out. Then I checked the landing gear circuit breaker to be certain that it hadn't popped and told my instructor that I would reduce the airspeed to below 87 knots and open the emergency gear release valve. My instructor frowned and gave me the checklist. "This is what you want," he said.

Apparently, before extending the gear in a Sierra, the pilot should not only make sure that the gear switch is in the down position, but also pull the circuit breaker. A big "WARNING" in the operating handbook advises that the gear malfunction may have been in the gear up circuit, and that with the breaker in, the gear may just decide to retract — on the ground. I was used to the emergency gear extension procedure for an Arrow, which requires that the pilot check to make sure that the circuit breaker is in.

Although checklists are the bible when it comes to malfunctions, some checklist items have to be committed to memory. In an engine-out situation, the first step is always to trim the airplane to best glide speed. Glide speed is a memory item, and so is trimming. In the event of an alternator failure, the first step is to shut off the master switch or to reduce the electrical load item by item. Of course different airplanes have different emergency requirements, but when time is critical, steps to increase glide distance or battery life ought to be automatic.

Another facet of learning the emergency procedures for an airplane involves learning something — as much as possible — about the airplane's systems. Although the operating handbooks for most single- and twin-engine airplanes avoid great detail on aircraft systems, they all have a section that generally describes the electrical system, the vacuum system, the hydraulic system, the fuel system, and various other instruments and options. I've found that although the handbook gives only a general outline of how the systems operate, that is enough for one to start to understand some of the malfunctions that can occur. Only one of my many instructors ever bothered to really concentrate on my airplane's systems. He quizzed me incessantly, and it really helped. It's true, a pilot can't pull over and look for glitches, but if he has some idea of how the systems work, it helps to take the mystery out of a malfunction, and can help the pilot take proper steps to deal with it. When I attended FlightSafety's Learjet ground school, we studied, in minute detail (I thought), each of the airplane's systems. Memorizing systems schematics didn't bear directly on the steps that we took to remedy the situation; the latter were often mechanical. But the whole task of reacting to an emergency is simpler, I found, knowing the probable causes of the problem and the reasons for the checklist steps. The same thing applies to a simple, single-engine piston airplane.

After learning the steps necessary to respond to an emergency, I think that practice is a pilot's number one concern. I have a hard time forcing myself to practice some emergencies — such as engine-out landings — whereas there are other emergencies, such as a vacuum failure, that I don't mind practicing at all. But one thing I do know is that my performance during an actual emergency is going to depend on the amount of time I've spent practicing — whether the emergency is an engine failure or a vacuum failure.

Partial-panel flying is a critical skill for any instrument pilot, especially if he flies a single. It's also one of the easiest to practice in day-to-day flying; it's just a matter of covering up some instruments. I've also seen pretty significant improvements in my full-panel flying after practicing partial panel. The exercise is important, because I learn to revise my scan to the primary instruments that I will be using. I also find that I develop a greater appreciation for the capabilities of each instrument. Whereas I may have been using the turn coordinator only to check and see that I am not slipping or skidding, it takes the place of half of my artificial horizon during a vacuum failure. Although I may have used the vertical speed indicator during descents and during approaches, it takes up the slack for the other half of the artificial horizon. And of course there is the good old whiskey compass. At the best of times, all I have ever been able to do with that monster is remember that left is less, right is more, and that a standard rate turn will give me three degrees a second. I've found that compensating for northerly turning error and acceleration error are way beyond my level of comprehension, much less skill, so I stick to timing my turns. The procedure works amazingly well. Partial-panel flying with any of the instruments covered can be a difficult proposition, but I know that I would be completely unprepared during an emergency without the practice. A set of instrument covers can be a worthwhile investment even for a pilot with a back-up vacuum pump, and if that's too much, business cards work just fine on some airplanes.

Unfortunately, not all emergencies lend themselves to practice in the airplane. Some are just too dangerous to practice with full realism. But something is better than nothing. I can get a pretty good idea of my performance during an engine failure from how effectively I set up a power-off approach to a field or the shore of a lake. When it comes to electrical failures, though, I generally just work through "what if" scenarios as I fly along. It is helpful to run through potentially sticky situations, but the experience resembles an analysis more than it does a real emergency. This is where simulators can be a tremendous help.

Although I have never lost a vacuum pump in instrument conditions, I'm pretty sure I've come as close to approximating it as any one can without living the experience. At an instrument refresher course offered by FlightSafety International, I got my first taste of complete visual, motion-based simulators, and my first taste of what it must really feel like to experience a serious emergency in flight. I found that although my partial-panel flying was pretty good, I followed the failing artificial horizon into a descending left turn as it went toes up. I found that I had to not only deal with the malfunctions, but also know the malfunction was happening. This was new.

Vacuum pump failures are particularly insidious. It all begins with a vaguely uneasy feeling. The instruments aren't doing exactly the same things. The artificial horizon is drifting downward while the vertical speed indicator shows a 200-feet-per-minute climb. "Which is right?" "What's wrong here?" are the first questions. A pilot's ability to interpret the conflicting information will decide whether he will have an opportunity to use the partial-panel skills he's been honing. I was able to pick up the errors in the artificial horizon and the DG quickly enough to recover from what was becoming a full-blown unusual attitude. But I was waiting for malfunctions. I had been getting one after the other, and I was on my toes. I'm glad now that I had the opportunity to see it first in the simulator.

Even the less realistic simulators without visuals or motion are excellent for practicing emergencies. I spent quite a bit of time in an AST simulator working on my multiengine rating, and besides getting very realistic practice feathering engines and shooting single-engine approaches, I also worked through a battery of other failures without the dead giveaway of my instructor pasting covers over the instruments or announcing that the electrical system had just failed. Above all, simulator failures force the pilot to determine whether he's really had a failure, whether some technical goof is making it seem as though he's had one, and, if so, what it is. I haven't seen a better way to safely simulate an emergency.

Even after learning the emergency procedures for the airplane that I'm flying and practicing the procedures in the airplane or in a simulator, I still think that emergencies are my greatest challenge in flying. Staying on the alert for malfunctions without becoming a paranoid wreck, reacting to malfunctions effectively while maintaining a cool head, knowing how to act deliberately without wasting time: such fine distinctions make a difference, especially in instrument conditions. Unfortunately, they are mighty difficult to develop and master. I guess that's where the practice comes in. What better way to keep a cool head, and to operate effectively when the red lights are flashing, than to know that whatever comes up, you've seen it and handled it before.

SECOND PART   Richard L. Collins

When I get a chance to fly one of the sophisticated simulators, I like to have as much stuff as possible busted. One of the most challenging approaches that I've shot was after a complete electrical failure in a Learjet simulator. That left me with the small two-inch artificial horizon operating and lighted on the left side. Patrick Bradley was flying copilot for me; he had a flashlight, and on his side had airspeed, altitude, and vertical speed. All engine power instrumentation was gone, the availability of flaps was questionable. The landing gear was already down. I was flying attitude and making small corrections based on Patrick's reading of the instruments on his side. We made it (with a simulated 1,500-foot ceiling), and before it was finished the realism of the situation had me sweating.

A simulator exercise like that is good exercise, but an education it does not make. It really can't even be counted as how you would do in an airplane in a similar situation. The "wham, bam, booms" just mean a little more when there is real air between your butt and the ground. Where a simulator is often harder to fly than an airplane, the comforting thought that the coffee machine is in the break room just 100 feet away takes away the edge. On the other hand, the simulator is infinitely more valuable than the airplane for training because there's no way anyone of sane mind would go out and practice real approaches with only the emergency equipment working.

The main thing is to take each unusual event — in the airplane or in the simulator — and grade performance on an individual basis. In a simulator, you just naturally expect the instructor to slowly (or suddenly) cripple the ship. On the other hand, we all spend mindless hours droning along in the airplane and putting a lot more faith than is justified in systems and other things. When something unexpected happens, we might be startled, which is a bad state of mind to be in as things come unraveled. Or, we might be overwhelmed with a dose of the right stuff and dismiss it as nothing more than a little old explosion in the engine compartment and try to carry on normally. The calm reaction is good, but procrastination can lead to trouble.

A relatively minor event illustrates the process when procrastination holds sway.

It was both a cold and a long day that had started in Las Vegas, where we know everyone gets a good night's sleep before flying the next day. The first hop was to Garden City, Kansas. After gassing there, I discovered some water in the left tank. It took a lot of draining, and I still used the water-free right tank for takeoff and climb — just in case.

Indianapolis was the next stop. It was near zero there, and I got a good case of the shivers while preflighting the airplane after it was serviced. I was paying special attention to draining the sumps because of the earlier water. I shivered and drained and shivered and drained. Then to illustrate the effect on thought processes of a long and cold day, I did the best I could to start the engine with the mixture in idle cutoff. That this wouldn't work was obvious, and I chided myself for not using the checklist properly and got it going. Time, I thought, to slow down and be more methodical.

Right before takeoff I removed my gloves and put them atop the instrument panel. Flying is something I just can't do with gloves.

As the airflow through the airplane increased on takeoff, the pilot in the right seat said he smelled gasoline. Hmm, at about liftoff, I too started smelling gasoline. Abort the takeoff? Certainly, in a 210, fuel does flow through the cabin en route to the engine. There are lines, the selector, and two reservoir tanks inside the cabin. I thought about that. If a gas cap is left off, fuel flowing out of the tank can often be smelled inside, but the cap usually bangs on the wing with a vengeance. I thought about that. By this time the procrastination made an abort academic. There wasn't enough runway left. Then I identified the source of the fuel odor. In my eager and shivering sump draining I had pretty well saturated my gloves, now atop the instrument panel, with fuel. As the airflow started through the cabin on takeoff the odor from them drifted back.

That's minor, sure, but I did go through the process of evaluating my response. I had one big question for myself. If it had been an IFR departure in low conditions (as opposed to an IFR departure in severe clear weather) would the smell of fuel during the takeoff have prompted an abort? I can only say that I hope so. In retrospect, it probably should have caused a stop for investigation even in good weather.

SIMPLE VERSUS SOPHISTICATED

The more complex the devices that we fly around with, the more chance of something breaking. It would be nice if that weren't so, but a machine with a million parts simply has a greater chance for a part to fail than does a machine with a hundred parts. So we have to be realistic and acknowledge that it becomes more difficult to complete *every* trip with *everything* working. We have to learn to evaluate glitches and to separate the ones that dictate an immediate change in plan from the ones that are a (usually expensive) nuisance.

SYSTEMS

Face it, systems in airplanes are not exceptionally reliable. Besides causing inconvenience, failures of systems in airplanes — in all types of airplanes — lead to accidents, as we have seen. I can recall three Part 25 transport category airplanes that have crashed in the past few years because of electrical system failures. And I have enough experience with systems in relatively sophisticated piston airplanes to know that while a system failure might evoke swearing from a pilot, the more logical response might be "Oh, it's you again."

As an example, at the 2,500-hour point on my P210 I calculated the mean time between vacuum failures to be 408 hours (it was much lower early in the operation of the airplane) and the mean time between charging system failures to be 306 hours. These systems are important enough and the failures frequent enough that I'd be a dolt to fly the airplane without always accepting the possibility of a system failing. (Lest this make the P210 sound unreliable, let me offer some other facts. In the 2,500 hours there were but two cancellations and one delay caused by mechanical problems other than vacuum or electrical. The cancellations were caused by a bad mag and a bad oil leak; the delay was caused by a recalcitrant cabin door. Of fifteen vacuum-alternator events, only four caused a flight cancellation. On the others, I was either inbound to home, could continue VFR, or was able to get the problem fixed with a minimum of delay. I do carry a spare pump and alternator. The result is that only a half of one percent of the proposed flights were canceled in 2,500 hours.)

There is no question that in either a single or a twin the

most common outages will relate to the vacuum or electrical systems, or instruments. Unfortunately our training airplanes are not equipped with a "failure" feature that the instructor can use to sneak up on our blind side. And we don't have sophisticated simulators that can be used to explore all the failure modes of particular airplanes. But we can think through all the failures and mentally simulate what will happen. There has been enough experience with failures and accidents related to failures to give a broad background on the subject.

QUIZ

First a quiz. On every airplane that you fly, do you know what instruments would be left after the failure of the vacuum system? What about after a failure of the electrical system? Do you know what autopilot functions would remain in each case? Do you know what autopilot functions would remain after the failure of any one instrument? Unfortunately, most pilot operating handbooks do not detail these items. It is left to us to determine the effect of various failures. If you can't figure it out yourself, ask a flight instructor, a person knowledgeable about the systems, or the avionics shop.

On most single-engine airplanes, the loss of a vacuum pump usually takes away the artificial horizon and directional gyro. If the autopilot is anything other than a basic wing leveler, it too will probably become inoperative if vacuum is lost. There are exceptions, and it is essential to know what the autopilot capabilities will be after a vacuum failure. Also, in better-equipped airplanes, some horizontal situation indicators are electrically powered, leaving this very useful instrument available after a vacuum failure.

When a vacuum pump quits, the gyros will slowly spin down and finally park. It has been my experience that the pitch indication on the artificial horizon is the first thing to show the effects of a system failure. And I have found it imperative to cover up the inoperative instruments. The artificial horizon usually parks in a 45-degree bank; this can be very disconcerting and can lead you to try to level the wings based on the information there. Covering the dead instruments also helps reorder the instrument scan.

THE REAL WORLD

The potential difficulty of understanding and dealing with failures that affect flight instruments is clearly outlined in an NTSB accident report covering a Lockheed Electra that broke up in flight one night a few years ago.

The crew knew they had a problem. A few minutes after takeoff they advised the controller that they had lost all electrical power and requested no-gyro vectors to visual conditions and an immediate descent to a lower altitude. The first mention of anything unusual came but one minute after the airplane was cleared for takeoff; the flight called for a radio check, and the controller said that the transmitter being used was better than the previous one. Then, four minutes later, the captain said to the first officer, "Ah, something wrong here with our VOR. . . . Yours appears to be right, mine's gotta be wrong. . . . Radar's not working for [expletive]. . . . My scope isn't [doing] nothing. . . . That's not bad. . . . Ya see, my picture looks like it's on 150 all the time."

The first officer agreed, and a couple of minutes after the captain mentioned the VOR problem he repeated that the radar was worthless. About a minute and half after that, there was the sound of a power interruption to the cockpit voice recorder. The recorder came right back on, and the following conversation ensued:

CAPTAIN: Lost everything, we lost a bus.

FIRST OFFICER: Lost a bus.

CAPTAIN: Which one?

FIRST OFFICER: I don't have any altimeter, I don't have any heading reference at all. Do you?

CAPTAIN: No.

FIRST OFFICER: Have any heading reference?

CAPTAIN: You got a flashlight?

FIRST OFFICER: Yeah, I got a flashlight. . . . Ah, we're boost out too. [This, in reference to the power-assisted flight controls.]

CAPTAIN: Huh?

FIRST OFFICER: We're boost out, aren't we?

CAPTAIN: Yeah, give us the boost.

FLIGHT ENGINEER: You got it.

CAPTAIN: No, get the boost off. . . . Go needle ball airspeed.

It was at this time that the crew advised the controller they had the electrical problem and would need assistance. Then the cockpit conversation resumed:

CAPTAIN: Give us gyro headings, we have no gyros.

UNIDENTIFIED: I'll keep an eye on the west. . . . You're in a left turn now.

FIRST OFFICER: No, you are in a right turn.

CAPTAIN: You help me, help me, help me fly the airplane.

FIRST OFFICER: Okay.

CAPTAIN: What have we got?

(At this time, sounds associated with increasing airspeed were recorded.)

FIRST OFFICER: Three zero zero on the heading.
CAPTAIN: We need a heading of a . . . 310.
CAPTAIN: Okay, help me, I'm climbing. . . . Okay. . . . Help me,
I'm, I'm . . . Needle ball airspeed.
FIRST OFFICER: I got you in a right turn.

The airplane broke up in flight six seconds later.

The NTSB found that the crew had probably lost one of the priority electrical buses. After this loss, the first officer's flight instruments, with the exception of the airspeed and the turn and bank indicator, would have been inoperative with appropriate flags displayed. The captain's artificial horizon and horizontal situation indicator would have been operative but would have displayed warning flags reflecting a failure of flight director functions. The integral lighting in the instruments also would have failed. So, while the captain probably did have the basic gyros operating, the flags and lack of lights might have led him to believe that nothing was working. (Aircraft of this type were not required to have a standby artificial horizon, as are jets.)

To illustrate the extent to which the captain became disoriented, the NTSB compared the "Okay, help me, I'm climbing" statement with the actual condition of the aircraft at that time, as shown by the flight recorder. When the captain apparently thought they were climbing, the aircraft was descending at a rate of 12,000 feet per minute, the airspeed was at 270 knots and increasing rapidly, and the g-forces were building.

The Board found that the cause of this accident was a progressive failure of the electrical system, leading to the disabling or erratic performance of some critical flight instruments and flight instrument lighting in night instrument meteorological conditions. As a result, the crew couldn't resolve the instrumentation anomalies to determine proper aircraft attitude reference, and became disoriented and lost control of the aircraft.

This accident is both graphic as well as illustrative of how difficult it can be, even for a three-man crew, to sort out and deal with a complex failure.

The airplanes that most of us fly are a lot simpler than a Lockheed Electra. Still, any failure that disables flight instruments when the airplane is in actual instrument conditions must be considered life-threatening until proven otherwise.

UNCERTAINTY

Looking at the Electra crew's problems, we see that uncertainty over what has failed can cause a very basic problem. This is not a simple thing to dismiss, even where you have good redundant systems. For example, if you have two artificial horizons and one starts to do a lazy roll to the right and the other stays level, how do you know which one is giving the correct information? Certainly if you look at nothing other than these two instruments you have only a 50-percent chance of guessing which one is correct. Those are bad odds when the stakes are high. Likewise, if the turn coordinator or turn and bank is suggesting one thing and the artificial horizon something else, you can't very well resolve the problem by looking just at those two instruments. Other sources must always be used to arbitrate a dispute between two instruments.

The requirement is to consider as many things as possible. In a single, the first place to go is the vacuum gauge. If it's on zero, then you know that the instruments driven by vacuum are the likely culprits. The chore then is to activate an alternate source of power to get them back on line, or to move to partial-panel instrument flying.

When going to partial panel the importance of covering

those instruments that are not operating can't be overemphasized. Any clues given by the dead instruments would be erroneous and disorienting, and it is easier to develop a new scan pattern with the things you don't want to look at out of sight. Again considering the Electra, the pilot might have had basic attitude and heading information, but with his flight director system inoperative he had flags, and perhaps other needles within the instruments giving erroneous indications. It could be a confusing situation even though the correct information was in there somewhere.

There are things to buy to minimize the effects of a vacuum system failure. Simple systems operate off the manifold of the engine and provide emergency power to instruments. (These don't work well on turbocharged airplanes.) Some companies offer electrically driven standby vacuum pumps. Electric instruments are also available. Certainly some backup is desirable on any airplane, more especially on a retractable, which is more demanding when being hand-flown on partial panel than is an airplane such as a Skyhawk.

HELP!

When a pilot does wind up with nothing but partial panel and a lot of clouds between the airplane and a friendly runway, the situation should be treated as an emergency. The controller should be told that there has been a loss of power to flight instruments. The pilot can transfer some of the work from the cockpit to the ground — navigation and information gathering — and can concentrate on keeping the wings of the aircraft level. A singlemindedness of purpose — flying — is a lot easier when you don't have to look up frequencies, look at charts, and write down sequence reports. Keeping the wings level is the clue.

When flying partial panel, beware changes. I was flying along with Patrick Bradley one day, in cloud, when he suggested that he'd like to try some partial panel in actual conditions. I cupped some business cards over the gyros, so that I could see them and he couldn't. He set out to manage the airplane in clouds with only a partial panel of instruments. Everything went well for a while; then the controller called with clearance to climb to a higher altitude. Bradley's control touch got jumpy and jerky, the heading started to wander, and I could tell that he was working very hard. While just droning along he had become quite comfortable with the task, but the change in the status quo was upsetting. It took him a while after leveling at the new altitude to settle back to the smooth job he had been doing before starting the climb.

VERY BASIC

The message is to settle in with some straight and level before trying anything fancy on partial panel. This might not always come naturally. When informed of an instrument or vacuum failure and faced with a pilot request to help out, most controllers would probably give the pilot a turn — unless the airplane happened to be pointed at the nearest airport. The pilot, though, should just stick with the ultimate in basics — straight and level — for a minute. Get the ducks in a neat row before proceeding. Even then, keep activity to a minimum. If one is in a position from which a straight-in approach could be made, that might be the best deal.

Where to go when something is busted? There have been

airplanes lost while descending from on top through clouds after a failure — even though they were within range of an area where a VFR descent could be made. Low approaches have been attempted on partial panel with bad results. Each case would have its special circumstances. If an approach has to be shot, the one with the best weather — preferably VFR — would probably be the safest.

A frequent result of instrument or instrument power-source failure is a loss of control. There's an important technique item here that is virtually always neglected. In a retractable, lower the landing gear; in any airplane, retard the power when loss of control is imminent or after control has been lost. Those actions put more time between the airplane and the ground. In the case of aerodynamically clean airplanes, they can help preclude an in-flight breakup caused by excessive speed in an out-of-control situation.

ELECTRICS

The Electra had an electrical system failure. It happens that it had electrically powered instruments, so the problem was much like a vacuum problem in a light airplane. Our electrical problems are more likely related to difficulty with the alternator or generator, and in most airplanes the relationship is more to avionics than to instruments. Electrical trouble has proven to be less hazardous — at least there are fewer fatal accidents as a result — but it can still be serious, especially if the failure of the charging system isn't caught quickly, before depletion of the battery.

For pilots flying other than local-area IFR operations, a battery-powered, handheld transceiver is an absolute requirement.

(If you'll run out and buy one, I'll quit harping on the subject.) If the airplane has an external antenna devoted to one of these radios, communications can go as well as they had with radios in the stack and will not be affected by anything failing on the airplane.

How might you use a handheld? I've worked mine a few times, usually in VFR conditions, but the examples are valid for IFR.

The alternator light came on one day just as I leveled off at 20,000 feet, westbound in my P210. The first consideration was the fact that I would have to leave the battery switch on to stay above 12,500 feet, where an altitude encoder is required. I thought I could do without that — I wanted to save the battery — so I got a clearance to a lower altitude. The battery switch was shut off; I knew that I had a good, fully charged battery waiting for whatever purpose I wanted to use it for. The weather was good on this particular day, so I got out the trusty map and navigated 250 nautical miles to a place where I knew that I could get the alternator fixed. It happened to be an airport with a TCA, so I turned the encoder back on when 40 miles out and used the handheld for all transmissions. The electricity was used for landing gear extension (which takes a lot) and for flaps extension and retraction, and the battery was still in good shape for starting after the alternator was repaired.

Switch the scenario to bad weather. I'd have done the same things except to use the handheld for constant communication with center to obtain vectors to an airport. Maybe I would have landed closer, but if weather or other factors had dictated a longer run I'd have been perfectly willing to fly 250 miles to get to a good place to land. Radar vectors can always be used for navigation — if you have communications capability. Should the controller lose the transponderless target, it is a simple matter to activate the transponder for a while to get reacquainted, and then turn it back off. When you reach the terminal area there should be plenty of power in the battery to crank up the necessities for an ILS approach.

Electrical load-shedding is important when the ship's battery is used after a charging-system failure. Everything that is not essential to the conduct of the flight must be turned off. Anything that uses much juice and doesn't have an on/off switch should have a pullable circuit breaker. With a good battery and judicious use of the energy stored in it, it's hard to imagine a scenario where you can't get down with little ado — especially if a handheld is along to help you conserve the battery.

When we fly with total reliance on the perfect functioning of everything in the airplane, any malfunction will cause an inordinate amount of trouble. And malfunctions will occur. Likewise, a cavalier attitude about things that are broken is dangerous. A recent airline accident appears to have been caused by electrical problems that intensified after the pilot of a twin departed with one alternator known to be inoperative and one suspect. Getting under way without everything going for you maximizes rather than minimizes the risk.

ONE SWITCH

There's no requirement of broken equipment for a pilot to get in trouble because instrument readings go awry. A Boeing jetliner apparently was lost because of a simple switch-flipping omission.

The aircraft, on a positioning flight, was climbing in an area where other flightcrews reported icing conditions. As the airplane climbed above 16,000 feet, both the indicated airspeed and rate of climb began to increase. The crew discussed the implications of this. The flight engineer suggested that it was caused by the aircraft's light operating weight. The captain said, "It gives up real fast. I wish I had my shoulder harness on, it's going to give up pretty soon." It was a turbulent night, and thunderstorms were forecast just to the south of where the aircraft was operating. Perhaps the captain was relating the climb to convective activity.

The indicated rate of climb eventually exceeded 6,500 feet per minute, and the sound of an overspeed warning horn was recorded as the airplane reached 23,000 feet. At that time, the recorded airspeed was 405 knots. The following conversation ensued on the flight deck:

CAPTAIN: Would you believe that [expletive].
FIRST OFFICER: I believe it, I just can't do anything about it.
CAPTAIN: No, just pull her back, let her climb.

The last comment was followed by the sound of the overspeed warning horn. The sound of the stick shaker (stall warning) followed on an intermittent basis with the recorded airspeed of 420 knots. The stall warning resumed and continued as the first officer said, "There's the Mach buffet, guess we'll have to pull it up." (The Mach buffet occurs when an airplane reaches its critical Mach number — the crew assumed this to be the source of vibration because of the indicated airspeed.) The captain commanded, "Pull it up."

The flight recorder showed that a couple of seconds later the vertical acceleration trace declined to 0.8 g (it is at 1.0 g in level unaccelerated flight) and the altitude trace started descending at a rate of 15,000 feet per minute. The aircraft was turning to the right. As the descent continued, the vertical acceleration trace increased to 1.5 g, the heading varied but moved basically to the right, and about 10 seconds after the descent started a crewmember transmitted to the air traffic controller, "Mayday, mayday . . . we're out of control, descending through 20,000 feet." When the controller asked the nature of the problem, a crewmember replied, "We're descending through 12, we're in a stall."

About five seconds after that transmission the captain commanded, "Flaps two . . ." The rate of descent remained the same, the vertical acceleration trace increased, the recorded airspeed decreased to zero, and the sound of the stall warning became intermittent.

Five seconds after the pilot asked for flaps the first officer said, "Pull now . . . pull, that's it." Ten seconds later the vertical acceleration peak values increased to more than 5 g. The rate of descent decreased slightly; however, the altitude continued to decrease — to ground level. The aircraft had descended from 24,800 feet in eighty-three seconds.

The cause of all this was rather simple. The airspeed and altitude values recorded were consistent with expected performance up to 16,000 feet. Then the recorded airspeeds correlated within 5 percent of indicated airspeeds that would be expected if the pressure measured in the pitot system had remained constant above 16,000 feet. Why?

The probable cause was listed as the loss of control because the flightcrew failed to recognize and correct the aircraft's high angle-of-attack, low-speed stall and its descending spiral. The stall was precipitated by the flightcrew's improper reaction to erroneous airspeed and Mach indications, which had resulted from a blockage of the pitot heads by atmospheric icing. Contrary to standard operational procedures, the flightcrew had not activated the pitot head heaters.

The relatively simple omission — not turning on the pitot

heat — robbed them of but one instrument, yet that one incorrect reading (with everything else apparently functioning) led them to pull the big Boeing up into a tremendous swoop as they tried to keep the airspeed indication on the proper value for a climb. The airplane would go only so high, it stalled, and they were unable to recover.

All that might sound obvious, but it is relatively easy to be fooled by something like this — especially when weather conditions are such that a big spike in rate-of-climb resulting from an updraft isn't the most unlikely thing in the world. And while this wasn't related to the failure of a system, it was related to the failure to use a system.

FIXATION

As we look at accidents in which experienced pilots lose control of airplanes, there's a strong suggestion that fixation on a problem, real or perceived, usually precedes such events.

A business jet was climbing out after takeoff from a major airport. The crew had been requested to expedite the climb through 15,000 feet, yet at 5,400 feet the climb was interrupted, and the aircraft descended back down to 5,000 feet. The controller asked the crew to verify that they were climbing to 15,000, only to be greeted by, "Standby one sir, please."

The NTSB felt that some unusual event must have captured the crew's attention, because at this point the aircraft started a turn back to the right where a left turn would have been required to get the aircraft on the assigned heading. Then the rate-of-climb increased dramatically, to a value far in excess of the airplane's capability in a steady-state climb. The crew didn't report a problem but was apparently absorbed by one and did not communicate for twenty seconds. Then they either didn't remember what the controller had asked or they hadn't understood it. They requested a "say again," after which the controller issued a new heading and asked their altitude. The copilot of the aircraft responded, "Okay, we're going up through about 93." The controller again issued the heading and the copilot acknowledged. This acknowledgment was made during an eighteen-second interval when the radar was not receiving the aircraft's transponder. The next transponder target indicated a change in track to the left and an altitude loss of 5,000 feet. A final target was received at an altitude of 2,100 feet about sixteen seconds after the 9,100-foot reception. The trajectory analysis indicated that the aircraft flew a steep-diving, left spiral, turned 270 degrees, and broke up in flight, at about 1,300 feet. The loss of altitude represented an overall linear velocity of about 413 knots and an average dive angle of − 39 degrees.

Clearly, control of the aircraft was lost. The NTSB's report on probably cause was "a failure or malfunction of an undetermined nature in the pilot's attitude indicating system, which led to a loss of control and overstress of the aircraft structure."

The airplanes we have discussed so far had crews of two or three. In every case it appears that a problem with the airplane was distracting to the entire crew. There is a lesson in this whether you fly alone or with another pilot.

Regardless of how bad things get, someone or something must keep the airplane in a safe flying attitude. As basic as that sounds, it is forgotten at times. When there is a diversion in an airplane, it is a time to be doubly wary. The airplane must be successfully flown if the problem is to be solved; for it to be flown, someone has to pay attention.

GURGLE GURGLE

I recall a distraction that taught me a lesson on this. It was back in the good old days of Narco Mark V radios that had both a panel unit and a black box in the back. The airplane, a 250 Comanche, had sat out in a driving rain the night before. I had gotten an IFR clearance on the phone before leaving an uncontrolled airport and was soon climbing out through the clouds. Things were okay until I started trying to communicate and make sense of navigational signals. Nothing worked, or was logical. I couldn't correlate the readings from two radios to establish a position, I wasn't in radar contact, and I was confused. Additionally, the communication radios were not working well; they could hardly hear me, and I couldn't hear them very well.

While trying to figure all this out, I went through a series of mini-unusual attitudes. I think the bank got to 45 degrees one time as I devoted 98 percent of my attention to trying to figure out what was wrong with the radios. After about twenty minutes of this, and after total confusion on my and the controller's part about where the airplane might be, I did my first logical thinking of the day. I knew where I had departed from, I knew my approximate heading since departure, or at least the heading through which I was swooping and dipping, and I knew how long I had been flying. That meant that I was about "right here" and if I settled down and flew a heading of 120 degrees, things would get better instead of worse. Forget the radio confusion for a moment. Then I had another good idea. I tuned the ADF to a strong standard-broadcast station about

100 miles ahead and had a reference to my geographical location. Communications improved. I made peace with the controller, flew into better weather, and finally learned why I had been so confused: The radios were soaking wet and highly unreliable. The airplane usually stayed in a hangar. This was its first time out in a very heavy rain. Numerous leaks around the windshield, and a very bad leak around the fresh air scoop that just happened to be located right above the black boxes in back, had doused the radios. I was ashamed of my performance. Embarrassing at best, but a good example of what happens when a pilot is distracted and pays attention to something other than the flying.

OUT OF BALANCE

In a twin turboprop Cheyenne accident, a couple of experienced pilots found that there's no requirement for something to break to cause a loss of control. In this case, all seats in the airplane were filled and the center of gravity was substantially aft of the limit for takeoff. As the aircraft climbed into clouds, control was lost. The way the airplane was loaded, it would have been longitudinally unstable and very difficult to control. There's a lesson here. In flight training, pilots are taught that the answer to problems is power. Power *on*. Stalls, the "loss of control" that is most often seen in flight training, are recovered from with power — lots of it. This must stick with pilots because in loss-of-control accidents the pilot seldom does anything with the power. It's "on," so that's good — but it can really be bad. In many loss-of-control accidents, a *reduction* of power might have helped matters. Power is destabilizing longitudinally; if the crew of that twin turboprop had reduced power to a value required to fly level, they might have been able to fly their improperly loaded airplane. If pilots of airplanes that break up in flight would pull power back to idle at the first sign of trouble, it could help.

SPEED

Speed is an important item when the airplane isn't doing well. It is not the crash that hurts, it's the sudden stop. The faster you are going, the more potentially destructive the stop. The fact that singles are more survivable in IFR premature arrivals shows how this works. Singles operate at lower speeds than twins on approach, so when they fly into the ground the stop is more likely to be survivable. That this is true outlines how we should fly when getting close to the ground in any emergency: at the slowest possible safe speed, or, usually, 1.3 times the stalling speed. A light airplane that is under control at touchdown, that doesn't collide directly with an immovable object, and that is going relatively slowly is a reasonably survivable place to be as long as the lap belt *and* shoulder harness are fastened.

Survivability is just one of the assets that we can count when stuff starts breaking on an airplane. Until something major breaks, such as the airframe, there is always a lot left with which to work toward a satisfactory solution. Maybe I'm like the kid digging through the pile of manure because he's confident that there is a pony in there somewhere, but I think a pilot who is prepared to make the most of any situation will manage.

Managing the pilot is as important as managing the airplane, too. The thought processes can get pretty complex in an emergency situation, but the thinking has to hold to what counts. I've seen pilots become quite excited over small things in airplanes — to the point that other things go to pot. Precipitation static that aces the radios causes some pilots to all but panic. Simple solution. Maintain the heading that was working, maintain the assigned altitude, and the static will go away sooner or later.

There have been a lot of cases of pilot's creating their own emergencies and then not being able to handle the self-induced problem. The Boeing with the pitot heat off is an example of that. It should remind us to ask the most important question when things go wrong: "Did I do something that screwed up the situation." Sometimes it is an omission, such as forgetting to turn on the pitot heat. Sometimes it is an action, like turning the fuel to "off" instead of to the desired tank. It's always amazing how many little glitches can be handled by just thinking back to the last thing done.

ICE VERSUS FLAPS

The analysis of a problem can be helped if you periodically spend time with the emergency section of the pilot's operating handbook. Imagine yourself in each situation. And learn from all of them. For example, in most Cessna POH emergency sections there's mention of restriction of flaps usage when there is ice on the airframe. There is good reason for this — a loss of elevator effectiveness can be experienced when there's ice on the leading edges of the flying surfaces and the flaps are extended. The admonition is valid for at least some other airplanes even though their manuals don't so state. But if you are shooting an approach some dark night with a load of ice and are contemplating full flaps because the runway is slick and you don't want to slide off the other end, a recollection of the relationship between flaps and ice might be helpful. It might make you consider the difference between diving into the ground beneath the point where flaps were extended and sliding off the end of the runway at 30 knots.

ICE ON THE REST

I like to think of all airframe icing as a glitch, akin to the slow failure of a system or some item of equipment. Ice isn't a sudden thing like a thunderstorm, in which you are either in or out, and except in extreme cases like the Colorado King Air, the effect of ice on the airplane comes about gradually. Like the illumination of an "alternator out" light, a zero reading on a vacuum gauge, or confusing readings from the engine or other instruments, the onset of airframe icing should be viewed as a signal that things have changed, that it's time for a new plan. That the action is dictated by a meteorological rather than an electrical or mechanical phenomenon is the only difference.

Ice might be compared with an alternator failure in a single because in both cases things will progressively worsen. There is only so much time in which to solve the problem. Just as there is only so much electricity in a battery to parlay into an uneventful landing, each airplane has the ability to carry a certain amount of ice before it loses the ability first to fly level and then to fly at all.

Ice is better than some of the mechanical things in one way:

The airplane came to the icing condition from ice-free air; that happy condition should still be back there awaiting your return. After takeoff, if ice starts forming, the good pilot reaction is the same as if a warning light illuminates: return and land. A little ice might affect an airplane only a little bit; let it build before taking the warning to heart and you are in progressively direr straits.

Earlier, Patrick Bradley related his first experience with ice, on his trip to Nashville, Tennessee, from Teterboro. When he explained the encounter to me after the flight, I started to ask him to reexamine his attitude about ice. There was a report, so it was *known* icing. He also picked up a little ice, so there was *known* icing. But when I listened to his rationalization, I was satisfied that the ice he was dealing with was, for example, like dealing with a flight with one of two nav-coms inoperative. That's when you make special conditions. Be wary of getting painted into a corner. It appeared to me that he was making the special considerations for a continuation and was sticking to them.

From a meteorological standpoint, we have to be certain when considering a "little" ice (as opposed to light, moderate, or severe icing) that this is all that's there. The general rule is that if you fly from no ice into some ice, you are probably flying into conditions conducive to ever heavier ice. That's the way it works. So there must be something there that is absolute proof that this won't be the case, if one is to continue. It has been my experience that the only time you can count on a successful encounter with a "little" ice is when the ceilings beneath are high, the tops are easily within reach and cloud layer is not too thick, and the climb or descent can be made through the clouds without hesitation. Only then is ice not an immediate threat.

THE ULTIMATE

It is fortunate that in general aviation airplanes, we don't often encounter any malfunction that is of necessity an immediate threat that can't be handled by remaining calm, paying attention, thinking, and acting. But a commuter airline Twin Otter crew gives us an inspiring lesson on how to manage in the worst possible scenario, when things unravel quickly into what is apparently a very life-threatening condition.

About fifteen minutes after takeoff, some ice was noted on the windshield of the airplane, and the first officer activated the deicer system, which utilized isopropyl alcohol. It didn't seem to be working, so he activated the system again and after a few seconds detected the odor of alcohol. He stopped the deicing. Shortly afterward, the crew saw smoke coming up through the floorboards.

Smoke in an airplane is very bad news, and, wouldn't you know, the controller didn't understand the pilot's first transmission. The crew repeated, and the controller issued a heading to fly to the nearest suitable airport. (When the crew first called, the last word in their transmission was "emergency." If it or "Mayday" had been the first word, the controller would no doubt have listened up and gotten the rest. This had no bearing on the outcome and is certainly not a criticism of a very talented and brave flightcrew. It's just a small point that we might as well learn from.)

By the time they got the airplane pointed in the proper direction, the smoke had become heavy and black, and it restricted the flightcrew's visibility to a couple of inches. They opened the side windows, but the visibility inside remained bad. The smoke was so thick that the first officer could not see the captain, but he felt the control input as the pilot started a descent. As the airplane descended below clouds (the ceiling at the airport toward which they were flying was 1,600 feet), fire broke out on the cockpit floor between the captain and first officer, as well as in the front of the cabin. A passenger who tried to put out the fire by covering it with his coat said that the first row of passenger seats in the airplane was in flames before the airplane landed.

The first officer reported that when he could see the ground he could tell that the aircraft was under control and that the aircraft was descending toward a frozen lake. The airplane impacted firmly on the ice. The left main gear and right wing and engine separated from the fuselage, where the fire continued. The two pilots and nine of ten passengers escaped from the wreckage; the passenger who didn't escape died of "asphyxia due to smoke, carbon monoxide, and hot-gas inhalation." Despite the flames one passenger returned to the cabin to rescue a teenage boy who was incapacitated.

During the in-flight fire, the captain and first officer sustained extensive second- and third-degree burns to their extremities, heads, faces, and upper and lower torsos. The captain's injuries were worse than the first officer's, yet he managed to concentrate on controlling the airplane and to make a successful off-airport forced landing despite the incredibly adverse conditions. From the report: "The Safety Board . . . commends the captain and first officer for their prompt and heroic actions in response to an in-flight emergency." We'd add, to Captain Thomas N. Prinster and First Officer Lyle W. Hogg, that all pilots owe you a debt of gratitude not only for the heroic actions but for setting an example. Of all the emergencies one might encounter, fire is probably the most distracting and most difficult with which to deal. Yet here are two pilots who were able to remain in control while everything went to hell in a bucket. It surely is a lot more rewarding to use examples where the crew worked things out than where the crew wasn't able to maintain control of the airplane.

If these examples teach us about handling an airplane after things break, this is the primary point. If the crews of any of the airplanes lost were able to read the accident report, they'd probably acknowledge that the situation should have been manageable, and, confronted with the same situation again, they'd handle it successfully. Given that, the rest of us learn that the keys are knowledge of the airplane and its systems, proficiency, and, perhaps most of all, a primary dedication to keeping the airplane under control while any problems are solved.

CHAPTER 9

Blowing the Whistles and Ringing the Bells

FIRST PART   Patrick E. Bradley

Sophisticated equipment — for navigation, weather avoidance, more accurate attitude information, and greater aircraft performance — has shrunk enough in size and price to whet the appetites of pilots that fly single-engine piston airplanes. And instrument pilots are the most sorely tempted by the opportunity to fly airplanes equipped with the latest and the best. I know, because I'm a prime example. Even with a limited flying budget, I generally won't hesitate to pay more to rent an airplane equipped with a few extra whistles and bells. Although much of my instrument flying has been in capable but minimally equipped airplanes, I have tasted the good life, and know that equipment can dramatically improve the quality of a flight. I also know, though, that most equipment will not enable me to complete flights that I otherwise would have had to cancel. Equipment just doesn't make you a better pilot.

But some equipment makes it easier to be a safer pilot. My first choice in optional equipment is a DME. It is simple, reliable, and a real work saver. By providing constant distance and speed calculations, the pilot can spend his time on the rest

236

of the cockpit chores. Moreover, a DME can calculate distance and speed far more precisely than I ever could; it offers nearly instantaneous information, given a good signal. With accurate speed information, it's simpler to calculate fuel consumption, and distance information is particularly helpful in locating intersections and determining when to start prodding the ATC man for a lower altitude. For instrument flying, a DME is the greatest thing since sliced bread.

For a long time, I was much less convinced of the merits of RNAV, but that was before I actually used one. To be able to place a VOR/DME at nearly any point your heart desires is a great benefit, especially during approaches. Although flying direct is often a convenient option, having the ability to place a VOR/DME on an airport can really simplify flying the approach more precisely and safely. Distance to the airport or the marker is there at a glance. Also, many airports have published RNAV approaches to runways not equipped with an ILS, VOR, or NDB approach.

Another area where pilots like to beef up their instrumentation is weather avoidance. I've had the opportunity to use radar alone and in tandem with a Stormscope, and in both situations I always end the flight with mixed feelings. Where some pilots tend to put implicit faith in radar to help them around thunderstorms or through cells, I tend to err to the opposite extreme. I have a hard time trusting my interpretation of the radar presentation. I'm never quite sure whether the clear spot is an anomaly or a bona fide clear area. More often than not, I think that I would be tempted to skirt the entire thunderstorm area, or go down low, where I could either see clear openings or make a quick landing. I could do that without radar. I have a bit more confidence in Stormscopes because I find them easier to read. Still, I'm not sure that I would want to use it for anything more than to determine whether I should divert right or left.

My problem with weather avoidance gear, then, isn't so much with the equipment as with my own ability to interpret it. I guess the answer to this is to learn how to use the equipment to augment, rather than to replace, the information on which I normally rely. I don't know, though, whether I would fall into the trap of believing the radar over my senses and ATC if the radar happened to be telling me what I wanted to hear at the time. Obviously, it takes a fair amount of experience, lots of savvy, and a heap of common sense. One other consideration with radar is that some models are more believable than others. A color radar is easier to read than a "green and white" model, and a radar with a ten-inch stabilized antenna will give more reliable information than one where the antenna is relegated to the inside of the leading edge of the wing; these could be an important consideration.

Upgrading the quality of the airplane's basic instrumentation can also be worthwhile for instrument pilots, especially the DG and the artificial horizon. Replacing the directional gyro with an HSI, slaved to a compass in the wingtip, ideally, simplifies flying approaches. With the nav information directly on the DG, it is simpler to see, at a glance, the relationship between the airplane's position, its heading, and the course. Also, having two instruments in one simplifies the instrument scan. Intercepting the localizer can be much simpler, too. I usually just keep my heading over the deviation indicator. As the airplane intercepts the localizer, I follow the moving CDI. Using this method, I can usually avoid overshooting or undershooting the localizer.

One instrument frequently overlooked in terms of quality that is worth serious consideration is the artificial horizon.

Some more complex airplanes are equipped with a better artificial horizon than we see in singles and light twins. Instead of a flat field, it is equipped with a ball that rotates behind the symbolic airplane. Precise attitude readings, down to a single degree if you're steady enough, are possible, and in instrument flying this can be a tremendous boon. I first used one of these in a FlightSafety Cessna 421 simulator, and I was amazed at how much the instrument simplified an approach. After setting the speed and establishing the proper descent rate on the glideslope, all I had to do was maintain my set attitude, and the airplane maintained its heading and descent. When I had to correct, I could accurately pitch up or down a hair's breadth and hold it. Technically, any artificial horizon should work the same way, but the difference in accuracy was almost startling. I came to depend on the near faultless precision. Although upgrading instruments can simplify flying, I think that I first might opt to install backup electric instruments or an auxiliary power source for the primary instruments. Accuracy and convenience are important, but the security of knowing that I will be in nearly as good shape if the vacuum pump quits as if it keeps inhaling is a comforting thought on a dark, foggy night.

I guess that the natural outgrowth of the black boxes that simplify flying is the black box that actually does the flying. Some highly accomplished pilots I know rely heavily on autopilots from the moment the airplane leaves the ground to the time the airplane reaches minimums on the ILS approach. In fact, I learned at Learjet school that the autopilot is an integral element of standard operating procedure. When flying with an autopilot the pilot delegates part of the flying chores to his mechanical copilot while he monitors instruments, checks the weather, and deals with the constant stream of decisions that are part of any IFR flight. The autopilot becomes a crew member, and just as the captain is responsible for monitoring the actions of his copilot, the pilot monitors and stands ready to take over for the autopilot in the event of any malfunction. The concept is reasonable, and it's effective. But why then, I ask myself, am *I* so reticent to delegate flying to an autopilot?

Part of the reason is because I'm not convinced that, at my stage of instrument flying experience, I could be certain of maintaining my own level of expertise without flying all of the flights myself. Sure, I'm glad to have an autopilot hold a heading for me while I study the approach to an unfamiliar airport, or while I'm copying the report of the weather en route, but I'm not yet ready to give valuable practice time to an autopilot that's already a better pilot than I am. And above all, I never want to get to the point where I rely on the autopilot to fly approaches as my own skills rust away. So, although I'm not sure that I'll *never* be an adherent of the autopilot-crew concept, for now, at least, I'll take the controls, thank you.

Optional equipment has come to play a key role in most pilots' instrument flying, and my feeling is that any instrument that can help me be a more accurate, safer pilot is a welcome passenger in my airplane. But too much of a good thing, in terms of black boxes, can be no good at all. Stuffing an airplane with gadgets that go unused is an economic waste, and my feeling is that any instrument that doesn't carry it's own weight should be sent to the graveyard for dumb, expensive mistakes. As a renter I avoid, whenever possible, paying extra for instruments that I won't use. That's never happened to me yet, because rental airplanes are seldom lavishly equipped, but I still hold to the principle. I guess what it comes down to is that I never want an instrument that I never use, and I never want an instrument that I can't do without.

SECOND PART   Richard L. Collins

There are two schools of thought on the amount of equipment that is desirable in a general aviation IFR aircraft. One,
the very basic, calls for flying with what is necessary — a couple
of nav-coms, an ADF, a marker, a transponder, an encoder,
and perhaps a DME. A simple wing-leveler autopilot would
be on the list of almost all pilots. At the other end of the
spectrum comes the pilot who likes dual area navigation gear,
dual weather avoidance systems, dual instrumentation, a full
autopilot-flight director, complete deicing, turbocharging, and
the Lord only knows what else.

The pilot with only basic equipment will have the advantage
of much lower maintenance bills, to say nothing of the money
saved on equipment. But the heavily equipped pilot has strong
assets in redundancy and information. Both approaches are
okay as long as we recognize that equipment decisions can
affect the airplane's utility. This is a key to safer IFR operations;
the act of attempting to extract more utility from an airplane
than is actually there often leads to accidents.

Many of the whistles and bells decisions are personal matters, too. For example, in my rather lavishly equipped P210, I
flew from the beginning with a Cessna 300 autopilot — heading
hold and nav tracking only, no pitch. The autopilot would
barely fly the airplane. I used to kid that it was the world's
lowest-maintenance autopilot. I never had to get it fixed because I couldn't tell when it quit. But the autopilot would sort
of keep the wings level while I tended to something else, and
it would keep the wings more or less level after a vacuum
failure (because the wing-leveling function gets information

from the electric turn-coordinator). It thus served an important function for me, and was what I felt I truly needed. To me, having dual instrumentation (electric and vacuum), airborne weather radar and a Stormscope, area navigation, full deice, and things like that meant more than a full autopilot. So when I added something, it wasn't a better autopilot. My theory was that I could fly the airplane, but I didn't want to try to fly it inside a thunderstorm — thus the appropriate equipment for that came first. Also, I valued the flexibility afforded by the deicing equipment more than I valued a device that would fly the airplane more precisely than I. Maybe I'll have a full autopilot by the time this book is published, maybe I won't. I'm not antiautopilot; other things have just taken precedence.

AUTOPILOT

Having raised the question, I'll first discuss whether the use of a full-service autopilot is a good thing. How do you best utilize an autopilot in single-pilot IFR operation? Is Patrick correct in not letting the autopilot fly?

Certainly there is nothing *wrong* with using an autopilot. A good one can do a beautiful job of flying an airplane. It can let the pilot relax and assume the roles both of systems monitor and backup. An autopilot is a real copilot, and there is no question that using one as such is more efficient than leaving it dormant. There's no way the machine can be awakened by the human's failure, to dash in and salvage a bad situation, but the reverse is true. Just remember, smart as some autopilots are, and as smoothly as they fly, none have as good a brain as a pilot. So it's logical to let the pilot think and the autopilot fly. Remember, instrument flying is a thinking game.

The big question becomes keeping the pilot proficient when

the autopilot is allowed to fly most of the time. Some of this can be handled by just watching the autopilot fly — they do such a good job that you can actually learn from watching. For example, you don't often see a pilot who carefully watches an autopilot personally do a lot of swooping and dipping on an ILS approach when hand-flying is required. He has watched the autopilot do its smooth and subtle job of tying down the localizer and glideslope and *knows* that holding precise headings and rates of descent — and making relatively small corrections — are the things that get the job done. On the other hand, autopilots fly mechanically, and pilots who use an autopilot a lot tend to do likewise. They tend to emulate the usually brisk roll rates rather than to fly with a smooth touch — an area where the human pilot can beat an autopilot hands down.

If there is a critical time, it comes when the autopilot fails. This might be because of a loss of power such as vacuum, because of an instrument failure, or because of a failure in the autopilot itself. The complexity of autopilots is such that failures occur on a reasonably frequent basis, and often you lose more than just the autopilot; some instruments go too. The pilot maintains the ability to step in with both practice and some hand-flying on trips and during approaches. That mechanical ability is easier to maintain than is the ability to think about the right thing at the right time. Still, the pilot who feels that he'd be in deep trouble if the autopilot didn't work, is likely to be quite correct in his assessment.

IN THE BUMPS

One place where pilots don't use autopilots enough is in turbulence. There's probably no better place to let the machine work and the pilot monitor. The altitude hold should be left off, and if the normal pitch control is overcontrolling it too should be turned off, if possible. But the roll control can usually be left on; in loss-of-control accidents involving light airplanes, lateral control is lost first. In strong turbulence, the machine might function better than the pilot. The pilot might let the machine handle the turbulence while he watches, ready to take over if necessary. In many loss-of-control crashes there was an autopilot on board capable of handling the airplane, but it was turned off. The reason is the same reason pilots barf when riding through aerobatics but feel fine when they themselves are flying: It feels better to be "in control" when in a wild situation.

MEET THE CREW

Regardless of the sort of autopilot in the airplane, or the total lack thereof, the pilot flying IFR alone does best thinking in terms of having a crew. The crew is the combination of equipment in the panel and the ability to discuss things with one's self. The interface between the pilot's brain and all the available information gets the airplane where it is going safely.

Navigation is a big part of instrument flying. Some of the new navigational devices can do a lot to make IFR easier for the pilot. They are good crewmembers.

Area navigation equipment, for example, is wonderful en route as well as in terminal area operations. It can do more for you than can a copilot spending full time trying to interpret raw data.

I like to set the area nav on the outer marker when maneuvering for an ILS approach. It then gives a reading of the distance to go until the time to start the final descent. This takes pressure off me. It also gives the distance to go to the localizer interception (because it gives off-track deviation in miles — a quarter mile per dot in the approach mode and a mile per dot in the en route mode). To paint the lily you can add radar graphics, which give a map display of the selected navigational problem on the radar screen. I've had a King KWX 56 with graphics, and with this, for an ILS, I set both the airport and the outer locator to appear on the screen. Selecting the localizer course bearing then puts the localizer on the screen. The result is a picture of the terminal area, and until you fly with such a system you just don't know how easy doing so is, compared with using raw data only. You can and should keep constant track of your position, regardless of how little equipment is in the airplane; area nav makes it easier. Once you learn to use it, it's like having a navigator on board.

LORAN

Loran is another form of area navigation gear that works well and makes possible direct navigation over long distances. It also has very accurate and sensitive cross-track error resolution. You can tie down wind drift more quickly with Loran than with anything else. It gives your ground track angle; at a glance you can tell what track you are making good. And then there are the more exotic long-range navigation systems that are basically reserved for airplanes with two commas in the price.

One thing you can't do with any area or long-range navigation system is save a lot of money. The direct distances are just not that much shorter than airways, and where an airway does take a big dogleg the controllers will usually help you go direct even if you don't have area navigation gear. The area nav is an informational device that does its best work in reducing pilot work load. And that is worth real money.

A thing to beware of is total reliance on anything for precise position finding unless you are flying a flight-tested area-navigation instrument approach or IFR route with equipment approved for IFR flight. The crew of the Challenger that hit the mountain at Sun Valley might have been using their long-range nav equipment for information if not for guidance. An error of a mile or two is not a big deal en route; when flying in a relatively narrow valley looking for an airport in marginal conditions, it can be a very substantial error.

PAY ATTENTION

On the other hand, don't ignore information that can be gained from area navigation equipment.

One place that I cherish the stuff is at Kerrville, Texas, a frequent port of call. Kerrville has a localizer (no glideslope) approach to Runway 30. It is usually a tough approach, because if it's cloudy in Kerrville it's usually right at minimums. And when it's cloudy, there's usually a decreasing tailwind on the approach and a slight downwind component for landing. It is an approach that'll work you pretty hard.

Area nav helps by defining the distance to go. There is an area nav approach to Kerrville (in the other direction) which is mighty handy, too, so there is a published and flight-tested waypoint on the far end of the runway. I set this up and know that when the area nav says I'm 1.8 from that waypoint, I'm about a mile from the approach end of the runway I'm going to land on. If the runway isn't in sight, there'll be no landing at Kerrville out of that approach. I'd use area nav in this manner only in conjunction with a flight-tested waypoint. The temptation is strong to use it for distance to go on all approaches, and that might be okay if you convince yourself that it's very approximate information. The height of folly would be to leave the MDA based on area nav information and without the runway in sight.

You surely can fly all the IFR you want without any form of area navigation gear, but the extra information that it provides is a real helper in the cockpit.

DME

The DME might be considered basic, but it isn't in a lot of IFR airplanes. Knowing the distance helps keep the mind straight, but I think DME's best value is in fuel planning. When things are getting close, a DME gives early warning of deteriorating groundspeed. It is much quicker than using time and distance, because there you are restricted to making computations after passing known points, usually VOR stations, and those can be rather far apart.

I was flying along one day, watching the groundspeed decay when it should have been increasing according to the winds forecast. I happened to be using the area nav function at the time, with a nearby vortac offset to the destination airport. The DME was simultaneously showing distance, groundspeed, and time to the airport. This readout was located in close proximity to the clock showing elapsed time since takeoff. All I had to do was add the time to go to the elapsed time to see why I shouldn't stick with the original destination. I knew to a few minutes how long the airplane would fly (five plus thirty) at the power setting, and I knew how long it would take to fly to the alternate, where the weather was worse than forecast. The total of that time, plus the required forty-five minutes, started flirting with my five plus thirty on the fuel, so I landed early and bought fuel. All this could have been calculated in another way, but having the information on the panel was like having another crewmember along to do computations.

THE SEAMY SIDE

The help from an instrument panel can become so complete that getting checked out on the equipment takes as long as getting checked out in the airplane. The time for checkout should be taken, though, because there are times and places where unfamiliarity with a system can combine with other less-than-favorable factors to best a pilot. This is illustrated by an airline accident.

The crew of the flight and the airplane formerly had worked for Northeast Airlines, which subsequently merged into Delta, the airline they were flying for on this foggy day. When the airplane was in its previous livery, it had had a Collins flight director; after it became Delta, a Sperry flight director was fitted.

As the crew was shooting a low approach, there was considerable cockpit conversation about the flight director.

CAPTAIN: Get on it Joe, ah, Sid.

(The first officer was flying.)

FIRST OFFICER: This [expletive] command bar shows.

CAPTAIN: Yeah, that doesn't show much.

CAPTAIN: Going like a [expletive].

CAPTAIN: Okay, your localizer startin' to come back in now.

FIRST OFFICER: Okay . . . set my power up for me if I want it.

CAPTAIN: Okay, just fly the airplane. . . . You better go to raw data, I don't trust that thing. . . . Let's get back on course if ya can.

FIRST OFFICER: I just gotta get this back.

The airplane hit short of the runway.

According to the NTSB report, the VOR/LOC tracking mode of the flight director would be selected for normal use. So set, the flight director's roll command bar would command the lateral maneuvers necessary to capture and track the localizer. Concurrently the system would arm to capture the glideslope, but only if intercepted from beneath. The flight director had an APP (for approach) mode that would capture the glideslope from above or below.

The flight was above the glideslope when it intercepted the localizer. So if the flight director had been set on VOR/LOC it would not have commanded a glideslope intercept.

After the localizer interception, the NTSB said, the flightpath and the crewmember comments indicated that the pilots were having trouble maintaining lateral position on the localizer centerline. The first deviation from the localizer came *after* the captain's comment, "Get on it, Joe, ah, Sid." At the time this was said, the airplane was still above the glideslope, and the airspeed was in excess of the proper value. The Board felt this could have prompted a switch in the flight director mode from VOR/LOC to APP. However, the subsequent steering problems would have been understandable only if the flight director had been placed in the go-around mode instead. This mode removes the localizer, and causes the flight director to command the action necessary to keep the wings level. Conceivably the first officer could have been confused by the pitch command displayed at that time. Certainly the conversation reflected the realization that the flight director wasn't providing reliable localizer or glideslope direction. Furthermore, the flight director was found to be in the go-around position after the crash, and there was no indication that they had intended to execute a missed approach.

Why the possible inadvertent selection of an incorrect mode for what they wanted to do? Well, with the system that had been in the airplane when it and the crew were flying the Northeast flag, a full clockwise rotation of the flight director mode switch would get you to "APP." With the system installed when they crashed, that action selects the go-around mode. The NTSB theorized that it is conceivable that a crew might, by habit, inadvertently select the wrong mode.

In an earlier chapter we explored the role of distractions in accidents. This crew certainly was distracted. Not only was the electronic help on the panel contributing to confusion, the NTSB, in its probable cause, mentioned poor positioning of the flight for the approach and nonstandard air traffic control services.

Remember the rule of thumb: When something is awry, ask the most important question: "What did I do to screw it up." Also, a low approach is a bad place to be dealing with equipment that isn't working properly, for whatever reason.

TURBOCHARGING

Turbocharging is very definitely an IFR-related option. It has been available for years. In switching from turbocharged airplanes back to normally aspirated ones, and vice versa, I've given a lot of thought to what might be the best deal for the pilot looking for a balance between IFR utility and cost.

For a while I felt strongly that turbocharging was a good IFR accessory bargain only for pilots flying in the mountainous west. There's so little real use for it in the east that nobody should bother, or so I thought. Most people don't or won't use oxygen. All turbocharging does is tempt pilots to fly at altitudes above 10,000, where without oxygen everyone suffers some effect of the decline in atmospheric pressure.

Then I switched my thinking a bit. After getting a pressurized airplane I decided that turbocharging was okay if accompanied by pressurization. Flying higher had its nice moments. Than I backed away from that a little and questioned in my own mind whether, in light airplanes in the 175- to 210-knot speed range, it made sense to have turbocharging under any circumstances.

After all that, I made a thorough survey of the way I used airplanes, and the benefits and costs of turbocharging — with or without pressurization. I realized I had been looking at the wrong things. I was trying to make it *pay*. Once past this foolishness, I decided that turbocharging is a solid-gold IFR asset, with pressurization if possible, but still of high value without.

The long-term costs of turbocharging and pressurization are real, but they have actually proven to be low in my operations. Nothing in either system seems to break, given reasonable care and a little maintenance from time to time. The overall costs are higher because of fuel burn, generally higher overhaul costs, and a somewhat more complex machine, but I feel the increase is reasonable. My P210 costs about twenty dollars more per hour to fly than would a normally aspirated 210. It saves little or no time, so the twenty dollars (about 16 percent) goes direct to cost. For it I get more comfort (the airplane is quieter and operates more smoothly because it has a heavier fuselage structure), I get some memorably fast eastbound trips at high altitude, but most of all I get the flexibility that allows the completion of trips that would be chancy in a lower-altitude airplane.

HOW SO?

On a wintry day I had a flight scheduled westbound across Pennsylvania, to the middle west. The weather was grungy at best, but flyable. The catch was in the icing forecast: "Moderate icing in clouds and in precipitation below 14,000 feet over the mountains." The winds were strong, about 45 knots at 12,000, and westerly. I'd have a headwind. Velocity increased with altitude, according to the forecast. There was some warm air near the surface along the east coast, but I didn't feel I could count on it being present inland.

If I had been flying a normally aspirated airplane, a V-35 Bonanza for example, the decision to go would have been a tough one. It would have to be predicated on climbing above any ice, which a Bonanza can do to some extent as long as the climb is started early. At the weight I'd have been flying a Bonanza (I was alone), it will climb 750 feet per minute at 12,000 feet and will go 500 feet per minute up to 16,000 feet. That's as well as I do in a cruise climb in a P210, but it'll maintain that steady pace to 23,000 where the Bonanza poops out rapidly above 16,000. I'd have to have oxygen on board the Bonanza, and the max cruise available at 16,000 feet would be about 160 knots. The only deicing available, for the prop, might help a little but not a lot. The question in the Bonanza would have been more related to the accuracy of the prediction that the ice would top out at 14,000. Given all factors, in a Bonanza I'd have made the decision to start out, looking back over my shoulder at the warm air down low over the coastal zone all the while. I could get high enough to be ice-free before getting far along over the mountains; failing that, I could tuck tail and return to base. I've done that before.

Go now to the planning and actual conduct of the flight in a turbocharged and pressurized 210. This airplane had the equipment required for flight into known icing conditions, but I've learned to plan flights to avoid ice where possible, regardless of the installed equipment. Those shiny boots shouldn't lead a pilot to feel that the airplane would win a full-scale tussle with substantial icing conditions. I wouldn't like to fly without it; deicing is essential on a high-flying airplane, because you can get into ice trouble in all seasons of the year and the deice does help clean up. But it is the altitude performance of the airplane that gives it flexibility in flight planning and in actual operation.

The plan was to fly at 14,000 feet, the top of the icing according to the forecast. That's the same thing I would have done in a Bonanza. When, in the actual flight, the P210 started accumulating ice at 14,000 feet, I turned on the prop deice and requested a climb to Flight Level 180. I could have gone to 220 if necessary. This was routine; there was no question about the airplane getting to FL 180, and from the outside air temperature at 14,000 (− 10°C) and the appearance of the clouds, I was pretty sure I'd be out of it at 180. Even in a Bonanza, one of the strongest-climbing normally aspirated airplanes, that ice at 14,000 would have been legitimate cause for buckets of sweat. Remember, once it starts it usually gets worse instead of better unless you do something to extricate the airplane from the icing condition.

Given the case, would you press on and try to climb above that ice in a Bonanza? Or would you accept the mistake in the forecast as a mandate to scratch the trip across the mountains and scurry back to the low-level warmth along the coast? Certainly prudence suggests that the latter course would be better.

That is an example of flight planning and operation where altitude capability helps with weather. It has done it for me enough times that I think the options available to a pilot increase as the square of altitude capability. Effectively doubling the vertical operating envelope with turbocharging quadruples the bag of tricks. Like the value in area nav, that's worth real money.

WESTWARD HO

Another thing that turbocharging does is open up IFR in the west. Where my buddies with normally aspirated airplanes have to go out of the way to fly the low-altitude route, I can usually go on up to FL 210 or 230 and fly straight across or out of the mountains. Where they might be at their virtual ceiling at the minimum en route altitude, a truly optionless existence, with a turbo there are four or five usable altitudes in each direction. You do have to take this capability in context, though. It doesn't repeal the law of gravity, and there are days when you can get into real trouble over the high mountains in an airplane with 23,000- or 25,000-foot capability. The deal in the mountains with turbocharging is about the same as the deal in the east without it.

The turbocharged airplanes also carry a strong requirement for better pilot knowledge about weather and the machinery. The engines are highly tweaked, and improper operation can ruin one quickly. Good engine instrumentation should be coupled with a studious pilot to take care of the expensive metal. If I had one bit of blanket advice on turbocharged engines, based on experience, it would be to run with a richer mixture in climb than is required (where leaning is permitted in climb). The desirability of this is recognized with some of the newer engines, where they just run full rich for a climb.

Turbocharging doesn't do much for speed because it's probably slower down low than a similar airplane without it. The better true airspeeds and eastbound tailwinds up high have to contend with long climbs so they don't really counter any low-level speed disadvantage. The turbo is a bit more complex, demands understanding and proper operating technique. In return, it gives far greater operational flexibility — a very nice thing to have in any IFR airplane.

WEATHER AVOIDANCE

Weather avoidance gear, weather radar, and/or a Stormscope go hand-in-hand with the additional capability offered by turbocharging. In fact, to try and operate a turbocharged airplane in its element without some form of weather avoidance gear can be quite risky. Flying in the middle levels does nothing to help storm avoidance. At 21,000 feet, for example, you are approaching the really mean storms at about their belly button. There, more than anywhere else, you need help.

Pilots who fly with weather avoidance equipment have fewer problems with thunderstorms, according to the accident records, and the equipment can enable a pilot to get more utility from any airplane without an increase in risk.

In thunderstorm *areas* is the key. There's not a thing in

the world that you can add to your airplane that enables it to operate safely in a thunderstorm. The stuff we buy is weather *avoidance* gear, and it must be used that way.

How do we use it? It's probably not as simple as some think. There is a lot more to intelligent use than glancing at a CRT and making a quick decision about how to handle a pictured situation. I will touch on the basics and encourage all to study further, attend seminars, and use the equipment conservatively.

The first requirement is an active interest in weather — meteorology if you will. To try and extract value from radar or a Stormscope without understanding the weather patterns that might be spawning thunderstorms is to shortchange the investment in the equipment. This curiosity about the weather system has to begin with the weather briefing and carry through to the completion of the flight.

A lot of emphasis is put on the outlook for severe activity. That can serve as a warning. If the TV or FSS folks are talking about a chance of severe thunderstorms, it's time to listen and stake out the areas where they think activity might occur. But in light airplanes we must be interested in all thunderstorms, not just severe thunderstorms. The response of our airplanes to what might be classified as a thundershower is enough to turn pink cheeks green and test the integrity of both the airframe and the pilot's skill. The only real difference between bad thunderstorms and worse ones (the only two kinds) is in avoidance distance. If it is a severe storm, it needs more berth. The disturbances in the air and clouds surrounding the cell itself can be strong a considerable distance away from observable precipitation or lightning. Thus there is a 20-mile guideline for avoidance of severe storms. Five miles is the guideline for garden-variety storms.

The weather pattern can define the type of storms with

which we will deal on a given flight. If there is an active cold front with real fireworks forecast, the strongest storms will likely be found in a squall line 100 or more miles ahead of the surface position of the cold front. Such a line might have passable spots, but finding one of them will require effort and perhaps a substantial detour. The key to using weather avoidance gear for this is to remember that you are looking for a *good* spot for passage, not the *best* spot. In a mean squall line the best spot might be impassable.

If there's a warm front, chances are thunderstorms will be imbedded in other clouds and in areas of rain. The bases might be rather high, allowing flight beneath the storms. Here the challenge is to fly in the rain and avoid the thunderstorms.

If it's a stationary front, the storms might be in short lines or they might be moving northeast along the front in clusters or clumps. Storms associated with stationary fronts are often (if not usually) obscured by other clouds.

In an occluded front (where in the most common scenario the circulation around the low is so strong that the cold front catches up with the warm front), the development of some horrendously mean storms is possible south of the low as and shortly after the occlusion occurs.

And when a low is developing rapidly, strong thunderstorm development might occur in the easterly quadrant.

Airmass storms, with no predominant feature on the map, might occur individually or in clusters. When there is just one, that's easy. When there are clusters or short lines, that is a sign that there is a weather system that wasn't dominant enough to make it onto the map. When we encounter more than isolated cells we should make the mental effort to think through what might have caused the outbreak. You can often look at the next morning's TV map and see what caused yesterday's problem.

STORMSCOPE

What are the primary assets of a Stormscope?

The WX-10 Stormscope plots electrical discharges on a CRT. The WX-8 displays the presence of them in sectors with a liquid crystal display. What any Stormscope tells you is that there is electrical activity in the indicated direction. Ranging is approximate, and it has been my experience that stronger storms appear closer and weaker ones farther away than the indicated range. There is also a characteristic called "radial spread" that extends indication of activity to closer ranges on the Stormscope display.

The way the indications come to appear on a Stormscope tells a lot about the situation.

The display can be cleared. If you get some indication, clear it and have activity start to come back quickly and actively, then pay attention to that sector — especially if the indication is showing up on a short-range mode or in one of the close-in sectors. By paying attention, I guess I mean turn the airplane so that it isn't pointing in the direction of the electrical activity. Clearing and watching both the timing and the severity of the return is a key to Stormscope use.

The Stormscope will indicate the presence and direction of a very strong collection of storms from hundreds of miles away. This has proven to be confusing to some pilots new to the Stormscope and has prompted requests for deviations around activity that is actually far beyond the destination. That's why the pilot's knowledge of the general weather situation is important. If you know there is a strong squall line in western Kansas, you know that it will show at the maximum range setting as you cross Missouri. To get rid of it and detect closer stuff, you'd utilize the closer sectors or the lower range settings.

The Stormscope can also appear to combine storms. If you have a moderate one ahead and to the left, and a strong one directly ahead but, say, 50 miles farther away, they might show as one. There's nothing wrong with that — the avoidance suggestion will keep you out of the storms. Just don't feel that the unit wasn't working properly in showing two as one.

The best summation of the Stormscope's capability is that its best work is in giving the big picture. It does not pinpoint individual cells or a collection of cells with precision. It gives an outline of where convective activity is present.

Here's an example of how this might be used. I was approaching Indianapolis one day for a fuel stop, with knowledge that there would be big weather problems on my next leg to Wichita. There was widely advertised severe thunderstorm activity over toward Saint Louis. Before landing at Indianapolis I studied the Stormscope display and knew that I would need to fly a heading of about 270 on departure, to stay on the north side of the activity. The picture was the same as I departed. The value of this was in my being able to aim for the north edge of the activity from a substantial distance away. The activity could not have been detected on my radar (with a ten-inch antenna) until the airplane was within about 80 miles of it. With radar, I might have flown toward the weather, gotten within about 40 or 50 miles, and then started trying to find a way through. Using the Stormscope as my primary information source, I was able to skirt the entire area.

RADAR

Airborne weather radar is more suited to the task of detecting and examining individual cells. The primary point of technique in using radar is learning to detect when the radar is attenuating, that is, when all the energy is being absorbed by precipitation and the range is effectively cut — to as little as 5 or 10 miles in extreme cases.

The primary way to check for attenuation is to tilt the radar antenna down to see if ground clutter is visible behind the precipitation return. If ground won't show, then you know there is attenuation. If it does show, then while there might be some attenuation, it is not absolute. That's nothing on which to bet everything. Again, a knowledge of the weather system is an important part of making a judgment based on what is shown on radar.

If you are approaching weather from the side toward which it is moving (usually the east or northeast) or from the side from which the cells are getting their feed of moisture (usually the southeast, south, or southwest), then what you see with the radar might be the worst that's there. There will be attenuation, surely, but it is not usually hiding any meaner stuff back on the other side. On the other hand, if you are approaching from the backside of a system, you might encounter an area of general rain that would obscure the radar picture of the leading edge.

A refresher: The next time a storm passes, watch the progression. If it's a good one, there will be a sudden onset of heavy rain and wind. The wind and rain will rage for a while, then both will begin to taper off. The taper might be gradual.

Certainly it will be gentle in relation to the way things got started. Now move back to the radar picture. If you look at the storm from the "gentle side," that gradual build-up to the climax on the other side will result in attenuation of the radar; it might not show the mean stuff clearly. From the other side, there's nothing between the radar and the worst of the storm to obscure the picture. If you have a Stormscope as well, it is unaffected by rain and attenuation.

I talked about actual use of radar in the chapter on en route flying; here let's look at some scenarios on how the use of radar might get us into trouble.

The most famous case involved an airline DC-9 approaching an area of severe weather from the back side. The airplane was operating in rain, assuring at least some attenuation. Additionally there was likely some ice on the radome, which also would obscure the radar picture. Approaching the most severe part of the storm system, looking for the best path, the crew made a decision to bear left. This turn actually took them toward the worst part of the storm; it appeared the best way to go because of attenuation. What might have looked like the last of the weather was actually the beginning of the worst of the weather. They suffered a double flameout in the heavy rain and hail, along with a shattered windshield.

A couple of accidents that I have studied seem related to the use of the radar tilt control. Tilt the radar antenna up too high and you might be looking over the top of the storm or at least at snow in the upper reaches of the storm. Snow makes a poor picture. (Hail shows poorly on radar, too.) To get a picture of the real gradient, you have to look at liquid; for best results, that means taking a cross section at a lower level in the storm. In a couple of accidents where an airplane lost a real battle with a squall line, the tilt control of the radar was found to be up, which would have given a false picture. Perhaps the moral is that if the picture looks bad, too bad to fly through, don't start fiddling with the radar controls in an attempt to make the picture look better.

THE PRIMARY POINT

Perhaps the primary point is that whatever is used to gain information on storms — traffic control radar, airborne radar, a Stormscope, visual means, voodoo, whatever — the name of the game is avoidance. Penetration of actual thunderstorm cells is not something that can successfully be done in light airplanes regardless of the level of equipment installed. Back in the earlier days of radar, there were some airplanes lost to thunderstorms. The word got out, though, and now airplanes that fly with weather avoidance equipment are seldom lost to storms, where we still lose about the same number of unequipped airplanes as always.

DEICING

Of all the things that can be added to an IFR airplane beyond the basics, deicing is probably the hardest one to justify by its actual use. It is available on most twins and on Cessna 210s and the Piper Malibu, and despite trouble with justification, if it is on the option list of an airplane I'd buy it every time.

The reason it is hard to justify is that it is so seldom used. When it is used, there usually is another course of action. If you have it, though, you might sit smugly at an icing altitude, cracking the stuff away when that really isn't the smart thing to do. You should flee any icing condition regardless of the equipment on the airplane. Where deicing is most helpful is in flight planning. "I don't think the ice will be there, but if it is, then I can use the deicing gear while I flee the ice." Even this might be a hazardous outlook. Every year the accident records prove that deiced airplanes (usually twins) crash more often in icing conditions than do airplanes without deicing. Most pilots without the equipment fly like cats — ready to jump at the first hint of something out of the ordinary. As a result, they have less trouble. I've used the analogy before, but it is always worth a repeat: Ice is like smoke in the cockpit. The first sign of it is a call to action.

STYLE

At the beginning of this chapter I mentioned that the extra things we put into an IFR airplane might be considered members of the crew. Once everything is there, then the pilot's job becomes the development of a management style that makes the most of the total package. If the style is not developed, then the real value of the equipment isn't realized. The autopilot is there to handle the rote process of flying, on your command. The area navigation equipment provides information about position. The DME gives quick and simple answers about the real effect of wind on groundspeed. Turbocharging puts into your throttle hand the ability to do more things with the airplane. Radar and a Stormscope give convective-weather information that might otherwise be unavailable. Deicing gives a means to at least partly clean ice off the airplane without landing. All those things make life easier and give a pilot

options to use in sometimes difficult situations. That brief syn-
opsis outlines the utility that can be *added* to a basic airplane.
None of it makes flying any simpler; it just makes it easier for
a good pilot to get the most out of the airplane. The pilot's
inward look at his ability is, as we will see in the next chapter,
the most important part of the equation.

CHAPTER 10

The Most Important Part

FIRST PART Patrick E. Bradley

"Watch it out there, you've still got a lot to learn." This caveat was the final bit of guidance my instrument instructor offered me before sending me off into the IFR world alone. I took the warning as a temporary limitation on my rating, like a student driver's permit. I thought that after flying some yet-to-be-determined number of approaches and getting an airplane wet a certain number of times, the limitation would be lifted, and finally I would have learned enough to be a full-fledged instrument pilot. I now guess that if anyone is going to lift that imaginary restriction from my rating it will have to be me, but with every IFR flight I take, I am more and more convinced that few, if any, pilots can ever honestly claim to know all there is to know about instrument flying. Not even the greatest virtuoso can claim to have mastered the piano. With each flight, there are new discoveries — nuances, subtleties — that expand my understanding of instrument flying. To assume at any given point that I can once and for all master the technique and all the variables of flying through the clouds would be dangerous self-delusion. I know that although I may be a good and safe pilot, there is *always* more to learn. I will always have to contend with certain limitations.

I get stuck on this last idea. Sure, I know it makes good

265

sense to recognize my limitations, but saying it is easier than knowing how far my skills will take me. And even if I did claim to know my limitations, how am I supposed to get any better if I don't push a bit further than what I think I can manage? I'm not sure that there are any answers to these questions. If there are, I'd be interested in hearing them. Until I do, I approach all my flights knowing that the time may come when I will have to draw the line. It all goes back to setting limits, trying them out, and setting new limits. I guess the important thing is to keep in mind that even the most accomplished pilots have their limits. I try to use my own judgment, being as objective as possible, other pilots' judgments, and some common sense to evaluate and reevaluate my limits.

I have yet to make a "perfect" IFR flight, one completely free of errors. I'm not convinced that such flights are completely necessary for me to be a safe instrument pilot. I do think it's essential to take stock of errors, and try to avoid them the next time around, or try to focus on weak links in my technique that I need to strengthen. In a sense, I try to make every flight a training flight. At the end of a flight, I usually have a pretty good feel for my weak points. And to ignore them would be to lose sight of my limitations. What's more, if I ignored them, I would run the risk of allowing my instrument skills to stagnate, or worse yet, to deteriorate. In many ways, my instructor's final words of advice to me are even more applicable now than when I was fresh out of my training program. It's easy to settle for mediocre but relatively safe performance. The only hitch is that someday performance might drift below "mediocre" and "relatively safe," and someday even relatively safe might not be enough.

For me, instrument flying is more than just a way to reach my destination in less than perfect weather conditions. Instrument flying offers a chance to take on challenges while developing a skill with practical importance. Sure, there are always disappointments. Finding that you can't fly an ILS as smoothly as you once could, or experiencing that vague, uneasy feeling that you're not as proficient as you were just a few months ago, are real setbacks. With practice and reasonable diligence, though, it's easy to turn the tables and move on to new challenges. Practice, simple practice, has been the key for me. If I'm having problems with a particular element of instrument flying, no solution is as effective as just setting aside time to go out and confront the problem. That way, I can concentrate more completely on what I'm doing, what I'm doing wrong, and how I should remedy the situation. Often, I know what I'm doing wrong, I just need the reinforcement of practice to get myself to do it right all the time. When it comes to approaches, I always need practice, even if I'm flying them well. If I let them go too long, I invariably lapse into bad habits, I just don't fly them as smoothly as I need to. I lose command.

I'm also a proponent of supporting local instructors. Even if I'm flying well, I can fly better if I go out with an instructor who can critique my skills, and wring me out a little bit. Instructors, good ones at least, have a way of uncovering hidden deficiencies. By challenging me, and pushing me further than I would normally want to or need to push myself, they can pinpoint where my skills begin to break down. When I see where problems lie, I can work them out with my instructor, and I can go on to practice them on my own. To have a good refresher with an instructor, though, it's pretty important, I think, to go in with an open mind and a flexible ego. No matter how well I'm flying, I know that my instructor is going to uncover flaws. Sometimes I've really got to work to overcome the urge to discount an instructor's unfavorable comments — "what does he know, I've been managing fine without his dumb advice" —

to get real benefit from what he's got to offer. A good attitude is one of the first requirements for improving as a pilot.

Evaluating and maintaining proficiency is complicated by switching airplanes, something that I do fairly frequently. Nearly all of my training was in a Cessna 172. After I got my rating, I checked out in an Archer, an Arrow, and after that a Sierra. Besides the differences in instrument layout and operating procedures in the different airplanes, I found that although I might be very proficient in one airplane, switching to another, even if I've been completely checked out in it, requires some transition time. Straightening out the operating procedures is one consideration, and switching from a slow airplane to a faster one is an even greater concern. Faster airplanes demand quicker reactions all around. I know that if I've been flying a slower, fixed-gear airplane, I might want to impose some limitations on myself when I switch back into a faster, complex airplane. Generally, the transition just requires a little refresher practice, so I try to make sure I get that practice before tackling the same conditions I would readily take on in the slower airplane. Transition to an even more complex airplane like a Cessna P210, with a turbocharged engine and pressurized cabin, requires yet more preparation. Although the FARs don't require that a pilot be current for every particular type of airplane he flies, I try to adhere to the common sense guidelines for my own safety. I just don't see any sense in striking out into IFR conditions, a taxing situation in any airplane, when I'm not completely on top of the particular airplane's basic flying characteristics.

Along the same lines, multiengine airplanes can pose proficiency traps for pilots who, although multiengine rated, spend most of their time flying single-engine airplanes. I recently got my multiengine rating and spent a majority of my training time honing my multiengine instrument skills. Flying IFR in a multi presents a whole new bundle of problems above and beyond just flying a faster and more complex airplane. Although the backup vacuum pump and generator would be improvements over most of the airplanes I fly, there is the nagging problem of how to react when one of the engines fails. The decision will depend largely on where, in the course of the flight, the failure occurs. Regardless, if there is a failure, a single-engine approach will be necessary. So when switching from a single-engine airplane to a twin, there is the added responsibility of being completely current not only in engine-out IFR procedures — responding to an engine failure in any phase of the IFR flight — but also on engine-out approaches. Staying sharp on multiengine IFR emergencies can take a fair amount of practice, and the temptation to gamble that there won't be an emergency is great. Statistics prove pretty conclusively, though, that multiengine pilots who don't maintain their engine-out skills, especially in instrument conditions, are in for big trouble.

Sometimes when I sit down and think about it, the responsibilities of maintaining an IFR rating can seem overwhelming. Staying current, staying in touch with instructors, and constantly pushing for improvement all take time, money, and a lot of hard work. Doing battle with the elements and short-circuiting the potential errors that seem to be waiting for the right moment to mar a good flight is a constant struggle. More than once I've asked whether, for a general aviation pilot like me, the effort is worth it. So far, I've had no trouble answering that question. Of course it's worth it. To be able to rely on an airplane as transportation, to be able to maintain what I see as an invaluable skill, is something special. If maintaining IFR flying skills is work — and it is — then it certainly isn't in vain. To be able to launch into the murk, fly along blue lines

on an en route chart, break out at 200 feet, and land at my
destination is a heady thrill. The work, in comparison, is a
trivial, and not all that unpleasant, requisite.

SECOND PART   Richard L. Collins

After looking at the various phases of IFR flying and at the
equipment, we come to the most important part of any instru-
ment-flying refresher. Self. The pilot in the mirror. To really
complete the job we have to probe our understanding of how
we approach the task at hand. Practice is important, but there
is more.

First, acknowledge that what we do for the FAA in getting
an instrument rating is but a beginning. The requirements,
as tough as they may seem to some, are superficial. Just as
most of us go far beyond what is required by law in equipment
for IFR operation, we have to do the same for the pilot. Couple
this with absolute honesty about how things are going and you
are well on the road to being a good pilot who understands
the risks and flies as safely as possible.

The airline safety record is far better than the record in
general aviation. We can learn something about a good way
to approach flying from the difference. Oh, airline pilots make
mistakes — witness the accidents used in this book for educational purposes — but they don't often do so. Why? It's
certainly not because they are super airmen. As a matter of
fact, most airline pilots are probably just average stick-and-
rudder pilots. A lot of them make lousy landings. But what
they do to insure success is to follow procedures in a methodical
way. This is insured by thorough training and proficiency
programs to make certain they don't stray far from the straight

and narrow. The procedures are backed up by meticulous flight planning and by use of challenge and response checklists. When an airline pilot does something illogical, such as run out of fuel, it can prompt a total review of that airline's operations and training methods.

This doesn't mean that airline pilots don't think, that they are thoughtless robots. Far from it. In fact, it means they have more time to think about the things that can't be covered by checklists and procedures. If the mechanical operation of the airplane and the instrument flying procedures are approached systematically, that frees the captain's brain to weigh the appearance of various shades of gray, to consider the report of poor braking action on the runway, or to mull forecasts that are going sour. Organization makes flying easier. If the old brain is in a state of disarray, flying is more difficult.

There is no question that transport-category airplanes make some contribution to the airline's better safety record, but look at it the other way. The only measurable number of accidents in general aviation IFR that might be eliminated if we all flew transport-category airplanes are the engine-out ones in light twins and the systems failures in singles. Those are a small part of the overall picture. We can do a lot more for the safety of our operations by flying and thinking like transport-category pilots than we could by flying transport-category airplanes.

UNDERSTANDING RISKS

In airline flying, pilots are not supposed to take risks. Is that statement true? Some might contend that it is, but airline crews, like general aviation pilots, have to do some things that involve more risks than other things. The risk in flying is lowest flying along in cruise on a calm, pretty day. The risk increases when the ceilings are low, when there is ice in the clouds, when there are thunderstorms about, when it is dark, when an organized storm system is working, or when something goes bump in the night (or in the daytime). The airline folks certainly don't fold their wings and park when the things that increase risk are in effect. What they do is work at managing the risks.

In an earlier chapter we explored the fact that low approaches at night, especially circling approaches, have proven to involve a very high, almost unacceptable risk. How might this be countered? For starters, there is a procedure for flying a night approach. If we do a good job of pilot briefing in the arrival phase of flight, that procedure is down pat. There is also a procedure to use if the runway isn't in sight at the minimum descent altitude. That's where we usually fail by descending a little extra for a better look. The procedure is there to keep the risk down to a dull roar; leaving the procedure can result in a loud crash. Why do we fail? If ever you bust minimums, get away with it, and don't have a little personal prayer meeting and swear off, you might as well be satisfied that a rock or tree out there somewhere has your name on it. Remember the airline accident in chapter 6? The one at New Haven? That crew that had descended below the MDA on approaches earlier in the day?

Think of busting minimums in relation to a simple thing like using a checklist. How many general aviation pilots do you know who use checklists? It's not an overwhelming majority by any means. But if we examine the differences between pilots who crash and pilots who don't crash, we might find that the ones who do crash are types who don't use checklists. It's looked on as a sissy thing to do, like missing that approach and having to go to an alternate. Any pilot who can't run his airplane without reading a list every time just need not apply to be a very hot pilot. Trouble is, VHPs are sometimes found to be extremely hot. Avgas and jet fuel burn rather well when spilled and sparked.

JUST DON'T

Where the risk involved in an approach and in not running a checklist are different, the pilot is accepting extra risk in both cases.

Some airlines and corporate operations address night circling approaches in a forceful way. Some have simply outlawed the maneuver. If a pilot does one he's in trouble with the company. Even though there is a procedure for a night circling approach, these folks don't think the procedure is viable. That's one way of reducing risk — by going beyond any FAA requirement.

Night approaches are infrequent events for most of us, but our approach to that checklist on *every* flight might give some insight into the personal style of risk acknowledgment and management. If the list isn't used every time, why? To save time? Because everyone knows you don't need such a list in simple airplanes? Because you never forget anything?

I took the Cessna checklist for my P210, a real hodgepodge that results in jumping from one place in the cockpit to another, and rearranged it so that the checking flows smoothly from right to left. That makes it easy to use. But I still go through dumb periods when I don't use the list. Instead, I go from right to left in a methodical manner, thinking that is just as good. And twice in a relatively short time I missed something and tried to start the engine with the mixture in idle cutoff. Any realistic self-appraisal would strongly suggest that not using the list is weakness, one that had better be addressed. If I forget the mixture, what next? The landing gear? The assigned altitude? Which way is up? It is a reflection of a basic sloppiness that creeps into the operation. Hereafter, you'll always find it used. I promise.

Our approach to the night business needs to be thought through before any night approach. "Here I am, it's dark, the weather is bad. This is where general aviation IFR pilots do themselves in. I am not going to flirt with the potentially fatal errors." Often, just the acknowledgment of being in a difficult situation is enough to preclude trouble. A pilot who has in advance gone through the possible pitfalls of a night approach is less likely to descend below minimums without a runway in sight or to descend into the ground with the runway visible in the distance. Saying to yourself that doing this might kill is a strong deterrent.

OTHER RISKS

A lot of things can give clues to the effectiveness of a risk management program. Anything that is done outside the law — however minor — is a bad sign. A lot of airplanes, especially twins, have a limited payload when filled with fuel. The airplane might have seven seats, but you can fill only three of them with full fuel. A lot of pilots overload these airplanes. We all know that there is a lot of pad in there, and perhaps think that the risk in flying over gross is small.

Realistically, the risk of flying in a slightly overweight condition *is* small in a single. But it's there, an additional risk, one that need not be taken. If the actual risk is small, the fact that the pilot is willing to take that risk is a big negative sign. It makes the event a sign of real danger. In a twin, it has been proven time and again that the overloading risk is small until something happens to one engine; then the risk becomes great indeed. In either case, it is the pilot's willingness to take risk for nothing more than the convenience of avoiding a fuel stop that is worrisome.

CONDITION

The condition of the pilot is another place where pilots take measurable risks. The medical folks give us all manner of guidelines — mostly involving "don't." This information is good, but I do think the admonitory nature of much aviation medical information prompts a lot of pilots to ignore it all. Pilot fitness is not a factor in a lot of accidents, and it has been logically argued that the third-class medical requirement could be done away with for private pilots with little or no effect on the safety record. But the way a pilot feels might be a causal factor in a lot of accidents.

Most pilots don't need a list of prohibited medicines, drugs, or physical conditions to know when they feel like flying. And if you hold up as a standard a day when you have had a perfect rest, have been eating a well-balanced diet, have been exercising regularly, and have no concerns, then we all take a risk in flying when we're in somewhat less than the optimum physical condition. Success is in evaluating any risk in advance. Certainly an experienced pilot flying on a clear VFR day with a little malaise resulting from a night out is not reflective of substantial risk. For an inexperienced pilot, that's not true. If it's a tough IFR flight, the experienced pilot might not fare so well. In the same condition, he might be on the verge of getting in over his head. The same might be true if the less-than-perfect feeling is from a slight cold.

A pilot who doesn't feel well might appear to do a perfect job of flying, to an observer, but only the pilot knows whether or not the condition is affecting his thought processes. Yes, we have to stay awake and be physically able to manage the controls of the airplane, but the fact remains that instrument flying is a thinking game; the effect on the thought processes is a primary concern when the pilot is feeling a little off the pace.

Booze and drugs show up in IFR accident reports occasionally. They are not a factor as often as in VFR, where related accidents often involve exhibitionism. And they certainly have no place in IFR because either affects our ability to think about the right thing at the right time. If ever we are tempted to break the eight-hour rule — even for one little beer — it is indicative of a deadly willingness to take risk. As for the residual effects of alcohol beyond eight hours, it's up to the pilot to decide when the ill-feeling no longer affects mind and body enough to interfere with the ability to fly the flight at hand.

In any area, a strong key is in the ability to decide not to take a risk. Even if all the other guys around the airport overload airplanes, fly after a few beers, complete low approaches when the ceiling is surely below the MDA, and land running on the fumes, to use a few examples, that doesn't make doing so right, or any less hazardous. Maybe most of them get away with it most of the time, but all don't get away with it all the time. The greatest waste is when the after-accident comment is, "Oh, [expletive], why did he do *that*?"

PREACHY, PREACHY

All that is pretty preachy stuff. It's too bad that it needs to be repeated, but it does — early and often.

The evangelism finished, it's time for school. At least it's time to think in terms of grading our efforts.

If we think in terms of just good and bad grades for flights, that's okay. But to get a more objective overview, and to quantify the approach to risk taking, it's enjoyable to try and put some value on grades.

We might say that the objective is to finish with low score, to fly the flight and accumulate no points at all — points being bad. It's like giving yourself a check ride, with points assessed based on the increase in risk caused by the infraction.

Things that get a lot of points, say, fifty, would be things that could in fact be life threatening. This would include obvious goofs such as starting a takeoff roll with nose-up trim left over from landing and neglected along with the checklist. Not tending to fuel and taking off with the tanks on zero might cost fifty. En route, diving into a dark and churning cloud with the thought that "I hope this works" would be good for fifty, as would failure to heed the first signs of ice. On approaches, the obvious worst case is a descent below minimums. Some hair-raising low-level maneuvering to salvage a botched approach would also qualify for high score. I'd also charge fifty for the failure to review the pitfalls of night approaches in the arrival phase of a night flight.

Next, to the twenty-five pointers. These would be things that were started or contemplated and that could have led to a life-threatening event. That they were rejected at the last minute gets the point score back to twenty-five. The critical preflight items caught at the last minute and out of sequence, the sweating palms that made you turn away from the storm before reaching it, or the night briefing that wasn't thought of until late in the proceedings are good examples of twenty-five pointers.

At ten points, we would come to the technique items. These are the things that are looked at on a check ride for the rating. That these items are far down on the point list is another example of how we have to go past the FAA requirements to stay out of trouble. Staying off altitude more than a hundred feet, wandering on the heading more than 10 degrees, not getting the approach radial or localizer and glideslope tied down early in the approach, putting communications ahead of flying, not getting the avionics properly set for the departure are examples of what I'd charge ten points for. If the failure to do any of these things leads to a more serious error, then charge twenty-five or fifty, as appropriate.

The least number of points, five, goes to the things that passengers are most likely to judge us on. Smoothness of flight and thumpy landings fall into this category. The quality of flight has to be near perfect not to get any five-point gigs.

That doesn't cover all possibilities by any means, but it outlines how you might give yourself a check ride on every flight. If each transgression is noted on the flight log as time allows, it will be remembered. You'll write down a lot less than a critical observer might, but if objective you'll start to get a good picture of your flying. It's fun to sit in front of the fireplace or by the pool and contemplate these things privately. *Why did I do that?* We all make mistakes. The only value in those mistakes comes when we study them.

FITNESS

There's big emphasis on fitness. This can be extended to flying. Just as physical fitness requires a plan, so does flying fitness.

The plan starts with the introspective look at our flying that we have just gone through. Then it moves to action. This comes in the form of making certain that when in and around the airplane, the necessary things are done in the proper order and that there's not a lot of attention, or thinking time, devoted to unrelated matters. Are we flying ambitiously, looking for a perfect score, or are we just trying to get there alive?

Allowing diversions causes trouble, as has been demonstrated many times.

An airline captain let his first officer and flight engineer switch seats before takeoff. In the preparations for takeoff, the crew apparently neglected to reset the pitch trim. Takeoff was with substantial nose-up trim. The flight engineer turned first officer was flying. The airplane pitched nose-up after liftoff, stalled, and crashed.

In another airline accident, the crew was discussing a wide range of unrelated subjects during the arrival and approach; the aircraft crashed on final, short of the airport.

In yet another, an overrun accident, the crew discussed unrelated items and was diverted by a member of the cabin crew during the approach. The airplane was high and fast, and the crew was unable to stop within the confines of the runway.

There's no way to keep all diversions away during the more

critical phases of flight. Passengers, air traffic controllers, and the machine itself might conspire to divert attention. A jumbo jet was once allowed to fly into the ground (on autopilot) while the crew attempted to resolve a problem with the landing gear indication.

Fitness is in answering any diversion with a double-up on the double checks. For example, if attention is forced elsewhere while running a checklist, there's an important question to ask when you come back to the list, "Am I starting where I left off?"

I was preflighting my airplane one day when a hangar neighbor strolled over to chat for a minute. After a pleasant interlude, I concluded the preflight and hopped on board. While running the checklist I rememberd that I was about to drain the sump on the left tank when interrupted. I had not drained it after the conversation. That's easy to fix in my airplane. You don't even have to get out because the left sump can be drained from the pilot's seat. I did just that and then slammed the door and went back to the checklist. It was just before takeoff that I realized that I hadn't fully secured the door. That was a first item on the checklist; I had done it and then undone it when I went after the sump. A minor item, perhaps, but illustrative of how diversions can foul up the thinking and checking process. The proper thing would have been to start over on the prestart checklist.

UH, DID I FORGET . . .

Letting uncertainty creep into IFR flying is a sign that a fitness program is being neglected. Have you ever, just when lifting off the runway, had a moment of uncertainty about whether or not you closed the hangar door? Or the baggage door? Or about whether or not your credit cards are in your wallet? I remember one day watching an otherwise very professional pilot make a hairy downwind landing because the passengers in his airplane realized at liftoff that they had forgotten something. They were on a tight schedule and had to have the items. Near panic must have followed. If he had run off the runway on that downwind landing, think how silly he would have felt about the whole thing. It was just one of those diversions. The really fit pilot has a system to check things so he can fly away confident that everything that can be done to insure a safe and uneventful flight has been done. When a diversion presents itself, it is handled in a calm and normal manner. Primary attention is given to managing the airplane.

GO FOR HELP

The talented instrument pilot has a good understanding of his relationship with the man on the ground. This understanding will include recognition that the man on the ground can offer some forms of assistance in times of high stress, but that basic responsibility for the operation of the airplane remains aloft, with the pilot. If ever there is a feeling that the successful completion of a flight depends on the air traffic controller, it's time for a reappraisal.

When you hear or read of a pilot saying something like "you gotta get me to the nearest airport," analyze the situation. If the pilot is asking for a heading to the nearest airport, he is asking for information. That's fine. The controller can help with information. But if the pilot has the feeling that the controller can actually get him anywhere, he's mistaken. In the

King Air with ice, in chapter 3, the crew explored several alternatives with the controller. They checked weather. They were far into their emergency when they made plain their plight, but then it was done rather clearly. Put yourself in a similar situation. If you had a problem soon after takeoff, knowing the weather at airports in the area would be a definite plus. And if it is a declared emergency, that makes the calling of all shots the prerogative of the pilot. "I am turning left and proceeding direct, descending to 8,000." The controller would then get everyone else out of the way or tell the pilot with the emergency that a serious conflict would develop as a result of the declared heading and altitude. Only the individual can judge his ability to make good decisions quickly in time of stress and then act on them. If your bad day should ever come, and we all hope that it doesn't, being able to think, decide, and act will be the key to getting out of that tight spot.

TRAINING

Training has a big role in maintaining fitness and is yet another area where we have to be totally honest with ourselves.

In its report on a light twin IFR accident, the NTSB called attention to the fact that, although he was properly rated, the pilot did not have much instrument flying experience in twins. Under the rules, the pilot had only to have a multiengine rating and an instrument rating to fly IFR in a twin. There's no requirement for any combination of the two. It might be argued that if a pilot is a good multiengine pilot and a good instrument pilot, there should be no problems inherent in a combination of the two. But if the complexity of the multiengine airplane results in one of those devilish diversions, the pilot's instrument flying ability might go right out the window. It is only logical that a pilot seek some multiengine IFR training or refresher flying before going off IFR in the more complicated airplane. The same is true of a simple-single pilot setting out IFR in a sophisticated single.

For training or refresher work to be of real value, it has to go beyond the bare requirements for the initial issuance of an instrument rating. And it should begin on the ground. Just hopping in an airplane and tearing around the sky is not the way to begin probing for weak spots in instrument flying. And before you throw yourself at the mercy of an instructor or refresher program, be satisfied that the people with whom you are dealing have actual experience in the type of flying that you do. There are a lot of instrument instructors out there who have little actual IFR experience and who probably don't want a lot of IFR experience in light airplanes. They are of precious little good to a pilot seeking serious refresher work. If the refresher is one to bring a rusty pilot back into the fold after a period of inactivity it might include some actual IFR flying.

Many pilots know little about the peculiarities of the particular airplane they fly. The training and refresher process is a good place to brush up on this. In a transcript of communications that preceded one accident, it was apparent that the pilot had lost the system that powered his flight instruments yet he seemed confused about what was working and what wasn't working. The same thing happened to the crew of the Electra that we discussed earlier. Some of the airplanes we fly might be complex, but they aren't so hopelessly complicated that we can't learn the systems.

Learning the systems: That has sold a lot of turboprop airplanes. To fly a pure jet, you have to go to school and get a type rating. Good, formal training. To fly a turboprop you don't have to have a type rating. Some pilots have stuck with turboprop as a result. They just don't want to go for that training. That's too bad, because the turboprops are as demanding as the small jets, and their worse safety record reflects the fact that formal training is not required.

I mentioned that flying as transport pilots do would improve general aviation IFR safety more than if we all flew transport-category airplanes. This includes doing what's necessary to insure that you could pass a type rating check for the airplane you fly, were one to be required. Besides being able to fly the airplane precisely and to handle all the emergencies, it means knowing the airplane and systems thoroughly. Where is the alternate static source control in the last airplane you flew IFR? How does it operate? If you instinctively reached for the correct location without hesitation, good. If not, knowledge of the airplane is lacking in that area, and probably in other areas. A moment of great embarrassment came to me once as I was departing in a twin that I thought I knew pretty well. With both engines turning, maps in lap, pencil poised to copy clearance, I started looking for the microphone. I couldn't find it and finally had to shut down and go ask someone if they had a spare that I could use. The reply was that if I'd reach to the right front of the pilot's seat I'd discover the new and exciting location for the microphone. If I had done my homework I would have asked someone if there were any changes in the new model.

Good training will bring precision to flying in general that helps when you are IFR as well as when you are VFR.

Take a simple item — the speeds used for various phases of flight. It's not uncommon for light-airplane pilots to think of speeds in a range; transport pilots think of them precisely. Proper speeds are calculated for takeoff and initial climb, based on aircraft weight, and the same is done for the approach and landing. Lifting off at a speed that has the same relationship to the stalling speed (which would decrease with weight) on every takeoff means that the airplane will respond the same. That's helpful when you fly IFR. In the jet training I've experienced, the instructors use one angle of bank when maneuvering, 30 degrees. Not 29 and not 31. In light airplanes, which go slower, a better value might be 20 degrees. Whichever, if you think in terms of the wings either being level or 20 degrees one way or the other, and of the speed on a calculated correct value, you'll see a new level of precision creep into your flying. It is possible to convince yourself of all this, but having it demonstrated by a good teacher is often more effective.

SIMULATORS

For those who fly turbine-powered airplanes and some of the fancier piston twins, simulator training is available. There's no substitute for this. The training value is wonderful. The rest of us have no simulators in the true sense of the word. At least, we don't have the kind that re-create our exact airplane electronically and offer a lifelike visual (nighttime) presentation and the motion that is calculated to feel as much as possible like the real thing. But we do have simpler simulators that can be used for realistic refreshing and training. We can also have as good a ground school on systems as the turbine folks, we can have as good a review on basic instrument flying, and we can take the desire to fly with precision to the airplane and give it a good workout there.

There is just no excuse for not being as sharp with your Skyhawk, for example, as the fellow in the blue blazer and tan slacks is with his Learjet. The difference is that you have to seek your refresher training and make sure it is tailored to your needs; the Learjet pilot is sought by training institutions, and the programs are neatly packaged for him.

PICKING THE AIRPLANE

The capabilities of various airplanes are cussed and discussed in a lot of places. The adaptability of most to IFR operation is obvious. The basic singles do fine but generally lack any systems redundancy. The cruising speeds are relatively low, meaning that a headwind can reduce progress by a substantial percentage. The range thus becomes a strong question, especially when alternates are elusive. On the other hand, these airplanes are very pleasant to fly, in instrument conditions. They are simple and involve the least possible work load. Step up a notch to retractables, and headwinds become less of a problem. But the airplanes are more demanding of flying skill. Thinking has to speed up to match the velocity of the airplane. With turbocharging, the pilot must develop knowledge of middle-level meteorology as well as come to grips with increased mechanical complexity. Pressurization adds a wonderful dimension to this category. Turboprops and jets offer very high performance along with amenities. From a pilot standpoint, the demands start to reach a peak with the turbocharged singles (if it is used to fly high, especially likely with pressurization) and increase slowly through the turboprops and jets. Yes, you read that correctly. There's almost as much to flying, say, a Piper Malibu as to flying a Citation. The primary difference is in speed.

The training and licensing system doesn't recognize this, and historically, when pilots have hopped into new high-flying singles without proper training and preparation, serious accidents have been the result. It has happened time and again. In some areas, such as proper engine operation, the pressurized piston airplanes require even more understanding than turboprops or jets.

As pilots, we have to tailor the airplane to our finances, our missions, and to the piloting ability that we wish to maintain. A big part of instrument flying is merging the airplane with the air traffic control system and the weather systems. It's a thinking game. That doesn't make IFR more or less equal in all airplanes. There's an additional dimension over VFR that increases the effect of airplane complexity.

Each IFR flight involves a certain number of actions and reactions. The faster airplane compresses all this into a shorter time frame. You have to think faster and react faster. That's not hard to do if you have a thorough understanding of the airplane and the air traffic control system. But it is a demand. A friend of mine who usually flies faster airplanes had a little IFR outing in a Cessna Skyhawk. He raved on about how much "easier" it was. The demand for thinking slowed down, and with each little heading or pitch variation on an ILS, the needles strayed at a much slower rate than he was used to.

Handling qualities are a factor. An airplane that has light controls might seem wonderful to fly on a clear day, but will work a pilot hard in clouds.

Altitude counts for a lot when you move out of the 10,000-and-below region. Besides the increased complexity of airplanes that fly high, when cruising up high there can be a lot of inclement weather between the airplane and the ground. Descents can be very interesting. During a bumpy descent into a busy terminal area, the work load can increase. It's a tough chore for a pilot who is used to thinking low and slow.

Those are a few factors. Historically, general aviation pilots have felt qualified to fly whatever airplane they can afford, with little special training. That is changing. The good old days of kicking the tires and lighting the fires are gone forever. At least, they are gone from instrument flying forever. There's too much at stake as we share the airways with an ever-increasing number of airliners, commuter and otherwise. It's perfectly fine for a person to want to remain a private pilot, with instrument rating, as long as the responsibility to fly like a professional pilot is met. It's far from impossible. It's not even difficult. All it takes is a good start — proper training — and a never-ending effort to maintain proficiency. Those are the marks of a pro.

TO EACH HIS OWN

Flying is a very diverse activity. Sailplanes, helicopters, seaplanes, aerobatic airplanes, sport airplanes, classics, antiques, ultralights, amphibians, and airplanes on skis provide great variety. The freedom of VFR flying is pretty wonderful. But of all the areas beyond basic flying, instrument flying has captured the fancy of more pilots than any other one thing. And there's as much satisfaction to completing an IFR flight in challenging weather as the aerobatic pilot might find in flying a perfect series of maneuvers or the pilot of an antique might find in the perfect landing on an idyllic grass strip. Instrument flying is demanding. To do it well requires thought and technique. It's not for everyone, by any means. A lot of pilots might choose from the other flying activities. But if you decide to fly IFR, and do it well, there are the rewards of getting where you want to be on schedule and of accepting and dealing with one of flying's more challenging tasks. Nothing, absolutely nothing in the world, turns me on the way a well-flown ILS approach to minimums does.

Index